EXPLORING MATHEMATICS ON YOUR OWN

Numbers: Their Personalities and Properties

Numbers: Their Personalities and Properties

C.D.H. COOPER

JOHN MURRAY · LONDON

Typeset by Tecprint Ltd, Loughborough, and
printed by offset in Great Britain
by Martin's Printing Works, Berwick-on-Tweed
0 7195 3097 0

Opening Remarks of the Chairman

'Good evening, Ladies and Gentlemen. I find it most gratifying that so many of you have turned out in such inclement weather to meet our little friends, the numbers. Too many people, I fear, make use of these dear creatures without stopping to think of each of them as an individual with his or her own unique personality. We are quite happy to see them overworked in electronic number factories but let us ask ourselves: have we ever stopped to talk to them as numbers? Yes, dear people, they are as different from one another as you and I and it is for this reason that I have brought some of them along with me tonight to tell you about themselves. Many other numbers are in the audience and later you will have an opportunity to meet them.

'I know that you're waiting eagerly to hear what they have to say, so I won't prolong this part of the proceedings any further and I will hand you over to number 0 to let her say a few words about herself.'

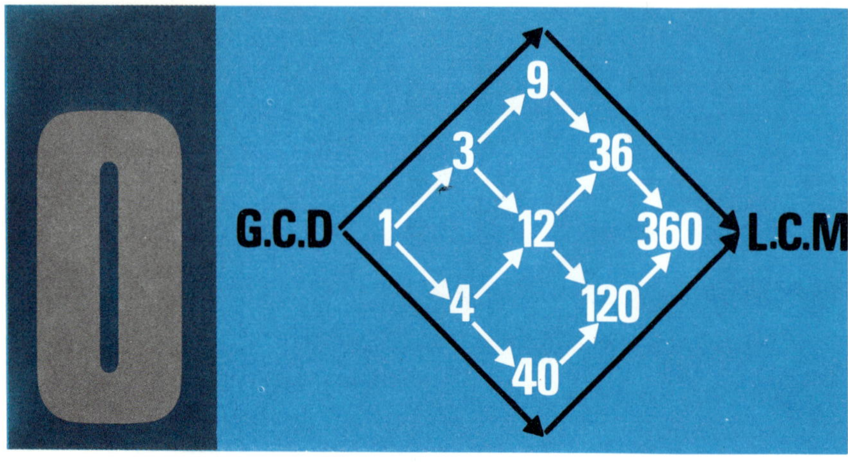

Divisors and Multiples

'Hullo. They call me 'naught'. I suppose that is because I'm not very clever. Somehow I don't seem to be very good at adding and multiplying. Whenever I add myself to another number it doesn't seem to make any difference and if I *multiply* another number the result is quite a disaster— the number disappears altogether and I'm left standing there by myself. I make such a mess of things they won't let me divide other numbers. 'You can't divide by 0,' they always say.

'I often wish I wasn't 0. Other numbers have to put up with only a few divisors while I have to put up with *every* number factorising me. One consolation I suppose is that those with no rights have no responsibilities. Other numbers have infinitely many multiples to worry about but I am my only multiple.'

Divisors

Throughout this book we shall use the word 'number' to mean one of the integers, or whole numbers:

$$\ldots\ldots\ldots, -3, -2, -1, 0, 1, 2, 3, \ldots\ldots\ldots$$

Number Theory is the science of numbers and is concerned with relationships between them. One such relationship—and certainly the most important—is the relationship of divisibility.

If three numbers m, n and q are related by the equation

$$n = mq,$$

we can describe the relationship between m and n in one of the following ways:

n is a **multiple** of m,

m is a **divisor** of n,

n is **divisible** by m,

m **divides** n,

m is a **factor** of n,

m **factorises** n.

These are just different ways of expressing the same relationship. For example, 15 is a multiple of 3 since $15 = 3 \times 5$. In other words, 3 divides 15, or 3 is a divisor (or factor) of 15. Another divisor of 15 is 5. Furthermore, since $15 = 15 \times 1$, we may conclude that 15 divides itself, and by writing this equation as $15 = 1 \times 15$ we see that 1 divides 15. Other divisors of 15 are -1, -3, -5 and -15.

Every number is divisible both by 1 and itself since the number n can be written as $n = n \times 1 = 1 \times n$. So 1 divides all numbers. On the other hand, 0 divides only itself or, put another way, 0 is its only multiple. This is because the product of any number and 0 is always 0.

The set, or collection, of positive divisors of a number n will be denoted by the symbol $D(n)$.

Examples

$D(0)$ = set of all positive numbers,

$D(1) = \{1\}$,

$D(2) = \{1,2\}$,

$D(3) = \{1,3\}$,

$D(4) = \{1,2,4\}$,

$D(5) = \{1,5\}$,

$D(6) = \{1,2,3,6\}$.

Notice that every number is divisible by 1 and itself. For some numbers, these are the only positive divisors. These numbers (apart from 1) are called *prime numbers* or, simply, *primes*. The number 1 is not included among the primes because of its very special properties. One could perhaps define a prime number to be one with exactly two positive divisors. This would rule out 1. From the above example we can see that 2, 3 and 5 are prime numbers. We will say more about prime numbers in Chapter 3.

Often we wish to observe that two numbers differ by a multiple of a certain number. If the numbers a and b differ by a multiple of m we say that 'a is *congruent* to b *modulo* m'. The number m is called the *modulus*. In symbols we write

$$a \equiv b \pmod{m}.$$

Examples. $15 \equiv 29 \pmod 7$ since the difference between 15 and 29 is 14 and 14 is a multiple of 7 (e.g. the 15th and 29th of a month occur on the same day of the week).

$32 \equiv 8 \pmod 3$ as 32 and 8 differ by 24, a multiple of 3.

All odd numbers are congruent to one another modulo 2 since any two odd numbers differ by a multiple of 2.

$n \equiv 0 \pmod m$ is yet another way of saying that m divides n.

We will consider congruence equations in more detail in the next two chapters.

1. Which of the following statements are true?
   (a)   3 divides 17
   (b)   5 divides 95
   (c)   4 is a multiple of 6
   (d)   8 is divisible by 4
   (e)   1 divides 371
   (f)   827 is divisible by 827
   (g)   91 is a factor of 7
   (h)   0 divides 13
   (i)   13 divides 0
   (j)   0 divides 0.

2. Write out the elements of $D(n)$ in each of the following cases:
   (a) $n = 15$,   (b) $n = 16$,   (c) $n = 17$,   (d) $n = 18$.

3. Which of the following numbers are primes?
$$17, 27, 37, 47, 57, 67, 77, 87, 97.$$

4. Investigate the claim that 'for all numbers n, $n^2 + n + 41$ is prime'.

5. In each of the following cases, choose a value of n from 0, 1, 2, 3, 4, 5, 6 so that the congruence equation holds.
   (a) $15 \equiv n \pmod 7$,   (b) $n \equiv 33 \pmod 7$,   (c) $83 \equiv n \pmod 5$,
   (d) $18 \equiv 37 \pmod n$,   (e) $2n \equiv 3 \pmod 7$,   (f) $n^2 \equiv 3 \pmod{13}$.

Greatest common divisors

The set of all *common positive divisors* of two numbers m and n is denoted by $D(m, n)$. These are all of the positive numbers which divide both m and n. This set will contain at least one element, the number 1. In the language of set theory, $D(m, n)$ is the intersection (or set of common elements) of the sets $D(m)$ and $D(n)$. Clearly $D(m, n)$ is always the same set as $D(n, m)$.

Examples. (The symbol ∩ denotes an intersection of sets)

D(4, 6) = D(4) ∩ D(6) = {1, 2, 4} ∩ {1, 2, 3, 6} = {1, 2} = D(2).
D(4, 4) = D(4) ∩ D(4) = D(4).
D(1, 3) = D(1) ∩ D(3) = {1} ∩ {1, 3} = {1} = D(1).
D(3, 6) = D(3) ∩ D(6) = {1, 3} ∩ {1, 2, 3, 6} = {1, 3} = D(3).
D(0, 5) = D(0) ∩ D(5) = {all positive numbers} ∩ {1, 5} = {1, 5} = D(5).

Observe that in all of the above cases, the set of positive common divisors turns out to be the set of positive divisors for a single number. Each D(m, n) is a D(r) for some r. Let us look at some more examples to see if this is always so.

Examples

D(18, 39)   = {1, 2, 3, 6, 9, 18} ∩ {1, 3, 13, 39} = {1, 3} = D(3).
D(150, 105) = {1, 2, 3, 5, 6, 10, 15, 25, 30, 50, 75, 150} ∩
              {1, 3, 5, 7, 15, 21, 35, 105} = {1, 3, 5, 15} = D(15).

It is beginning to look as though D(m, n) will always be equal to D(r) for some r. We shall prove that this must always be the case but before we do, can you see the relationship between r and m and n? Here are some more examples to consider.

Examples

   D(35, 8) = D(1),   D(24, 200) = D(8),   D(39, 65) = D(13).

In all cases, D(m, n) = D(r) where r is the *greatest common divisor* (g.c.d.) of m and n, that is, the largest number which divides them both. Sometimes it is called the highest common factor (h.c.f.) of the two numbers. The g.c.d. of m and n is usually denoted by (m, n). Thus, for example, the g.c.d. of 39 and 65 is 13 or, in symbols, (39, 65) = 13.

We have not yet proved that D(m, n) will always be D(r) for some r, but we can show at this stage that if this is so, then r must be the g.c.d. of m and n. For r is clearly the greatest divisor of r and so is the greatest number in D(r). If D(m, n) = D(r) then r is the greatest number in D(m, n), that is, the greatest common divisor.

Euclid's algorithm

The following algorithm (or computational process) appeared in the seventh book of Euclid's *Elements* in about 300 B.C. but it is highly probable that it was known even earlier. It is an algorithm for finding the greatest common divisor of two positive numbers. It depends on the fact that if a number d divides each of two numbers m and n then it divides both their sum and difference. (If $m = dq$ and $n = dr$ then $m + n = d(q + r)$ and $m - n = d(q - r)$, both being multiples of d.) Thus if d belongs to $D(m, n)$, then it is a divisor of m, n and hence of $m - n$. Hence it is a common divisor of n and $m - n$ and so belongs to $D(m - n, n)$. On the other hand, if d belongs to $D(m - n, n)$ it divides both $m - n$ and n and hence divides $(m - n) + n = m$, in which case it divides both m and n and so belongs to $D(m, n)$. We have therefore shown that

$$D(m, n) = D(m - n, n).$$

We now examine the way this simple fact can be used to find greatest common divisors.

Example. Find the g.c.d. of 91 and 161.

Solution. $D(161, 91) = D(161-91, 91) = D(70, 91)$. At this stage we can see by inspection that $D(70, 91) = \{1, 7\} = D(7)$, but let us continue the process instead.

$$\begin{aligned}
D(70, 91) &= D(91, 70) = D(91-70, 70) = D(21, 70) \\
&= D(70, 21) = D(49, 21) = D(28, 21) \quad = D(7, 21) \\
&= D(21, 7) \ = D(14, 7) \ = D(7, 7) \quad\ = D(0, 7) = D(7).
\end{aligned}$$

Thus the g.c.d. of 91 and 161 is 7.

The process we follow is to subtract the smaller number from the larger at each stage. Whenever the smaller number is not zero, this will reduce the sum of the two numbers. Obviously the process must stop eventually, and when this happens, the smaller number will be zero. We thus conclude that for some r,

$$D(m, n) = D(0, r) = D(r).$$

This proves the assertion made earlier.

The process can be speeded up somewhat by subtracting multiples of the smaller number from the larger at each stage. In the above example we subtracted 21 from 70 three times. We could have done this in a single step by subtracting 3 times 21 from 70.

We can generalise the equation $D(m, n) = D(m - n, n)$. For any numbers m, n and q,

$$D(m, n) = D(m - nq, n).$$

The proof is essentially the same as the one given above for the case $q = 1$.

At each stage in the algorithm we subtract some multiple of the smaller number from the larger. Which multiple? Obviously we want to decrease the numbers as quickly as possible and so we subtract the largest multiple which will leave a non-negative remainder. We therefore divide the smaller number into the larger and replace the larger by the remainder.

Euclid's algorithm

To find the g.c.d. of two positive numbers:

1. At each stage divide the smaller number into the larger and obtain the remainder.

2. If this remainder is 0, the smaller number at this stage is the required g.c.d.

3. If the remainder is positive, replace the larger number by this remainder and continue.

Example. Find the g.c.d. of 299 and 943.

Solution

$$\begin{array}{r} 3 \\ 299\overline{)943} \\ 897 \\ \hline 46 \end{array}$$

Thus $\qquad$ $D(943, 299) = D(46, 299).$

$$\begin{array}{r} 6 \\ 46\overline{)299} \\ 276 \\ \hline 23 \end{array}$$

Thus $\qquad$ $D(299, 46) = D(23, 46)$

$$\begin{array}{r} 2 \\ 23\overline{)46} \\ 46 \\ \hline 0 \end{array}$$

Thus $\qquad$ $D(46, 23) = D(0, 23) = D(23).$

Hence $\qquad$ $(943, 299) = 23.$

Here we continued the algorithm to the end, but we could have obtained the answer by inspection one step earlier. The g.c.d. of 23 and 46 is clearly 23. As the next example shows, the algorithm is very efficient and can obtain g.c.d.'s of quite large numbers without too much computation—certainly much less than if the factors of the two numbers had to be found.

Example. Find the g.c.d. of 318943 and 130217.

Solution

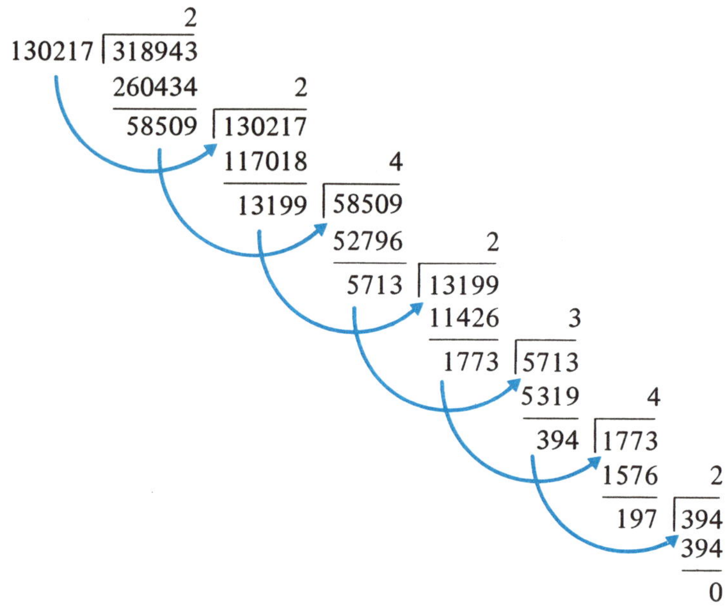

Thus $(318943, 130217) = 197.$

Two numbers are said to be *coprime* (or 'relatively prime') if their greatest common divisor is 1.

Example. Prove that 74329 and 3827 are coprime.

Solution

$$
\begin{array}{r}
19 \\
3827\,\overline{\smash{)}74329} \\
3827 \\
\hline
36059 \\
34443 \\
\hline
1616 \\
\end{array}
$$

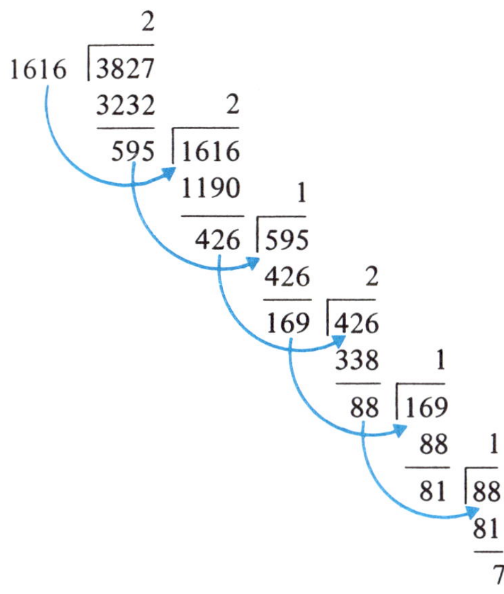

Thus $(3827, 74329) = (7, 81) = 1$ and so 3827 and 74329 are coprime.

EXERCISE SET 2

1. Find the g.c.d. of 3799 and 7337.

2. Find the g.c.d. of 123456789 and 987654321.

3. Are the numbers 1234321 and 4321234 coprime?

4. Prove that if $m - nq \geqslant 0$, $D(m, n) = D(m - nq, n)$.

5. (Harder) Find the g.c.d. of $3^{100} - 1$ and $3^{100} + 1$.

Least common multiples

The set of positive multiples of a number, n, will be denoted by the symbol $M(n)$.

Examples.

$M(0)$ = empty set (i.e. there are no positive multiples of 0),
$M(1)$ = set of all positive numbers,
$M(2) = \{2, 4, 6, 8, \dots\}$ = set of all even positive numbers,
$M(3) = \{3, 6, 9, 12, \dots\}$.

The set of all common multiples of m and n will be denoted by $M(m, n)$ and it is clearly the intersection of the sets $M(m)$ and $M(n)$.

Examples

$$M(3, 5) = \{3, 6, 9, 12, 15, 18, 21, 24, 27, 30, \ldots\} \cap$$
$$\{5, 10, 15, 20, 25, 30, \ldots\}$$
$$= \{15, 30, \ldots\} = M(15).$$
$$M(4, 6) = \{4, 8, 12, \ldots\} \cap \{6, 12, \ldots\} = \{12, \ldots\} = M(12).$$
$$M(3, 3) = M(3) \cap M(3) = M(3).$$
$$M(1, 2) = \{1, 2, 3, 4, \ldots\} \cap \{2, 4, 6, \ldots\} = \{2, 4, 6, \ldots\} = M(2).$$
$$M(0, 5) = \text{empty set} \cap M(5) = \text{empty set} = M(0).$$

As with sets of common divisors, $M(m, n) = M(r)$ for some r. Moreover since r is the least positive multiple of r, it is the least element of $M(m, n)$, that is, it is the least (positive) common multiple of m and n. We call it simply the l.c.m. of m and n.

G.c.d.'s and l.c.m.'s are closely related. For positive numbers m and n, the l.c.m. of m and n

$$= \frac{m\,n}{(m, n)}$$

Example. Find the l.c.m. of 24 and 30.

Solution. $(24, 30) = 6$. Thus the l.c.m. of 24 and 30 is

$$\frac{24 \times 30}{6} = 120.$$

G.c.d.'s and l.c.m.'s of three or more numbers

We define $D(n_1, n_2, \ldots, n_r)$ to be the set of positive common divisors of $n_1, n_2, \ldots, n_r$. It is the intersection of the sets $D(n_1)$, $D(n_2), \ldots, D(n_r)$. Similarly we define $M(n_1, n_2, \ldots, n_r)$ to be the set of positive common multiples of $n_1, n_2, \ldots, n_r$. It is the intersection of the sets $M(n_1), M(n_2), \ldots, M(n_r)$.

$$\textbf{Example} \qquad D(24, 30, 75) = D(24) \cap D(30) \cap D(75)$$
$$= D(6) \cap D(75)$$
$$= D(3).$$
$$M(24, 30, 75) = M(24) \cap M(30) \cap M(75)$$
$$= M(120) \cap M(75)$$
$$= M(3000).$$

Note that the above formula connecting l.c.m.'s with g.c.d.'s does not extend to more than two numbers. In this example,

$$\frac{24 \times 30 \times 75}{D(24,\ 30,\ 75)} = 18000$$

while the l.c.m. of 24, 30 and 75 is 3000.

Negative numbers

Although divisibility was defined for negative numbers as well as positive ones, we have concentrated in this chapter on positive numbers. Negative numbers do not, however, raise any special problems. The divisors of a number n are precisely the divisors of $-n$. For if $n = dq$ then $-n = d(-q)$. Hence

$$D(-n) = D(n).$$

Thus $D(m,\ n) = D(m) \cap D(n) = D(-m) \cap D(n) = D(-m,\ n)$
$$= D(-m) \cap D(-n) = D(-m,\ -n).$$

Thus the g.c.d. of m and n is equal to the g.c.d. of $-m$ and n and of $-m$ and $-n$. 'Minus signs' have no effect on g.c.d.'s. G.c.d.'s are always positive.

Example. Find the g.c.d. of -24 and 36.

Solution. $(-24,\ 36) = (24,\ 36) = 12$.

The multiples of a number n are precisely the multiples of $-n$. For $kn = (-k)\,(-n)$. Hence

$$M(-n) = M(n).$$

Thus, as for divisors, $M(m,\ n) = M(-m,\ n) = M(-m,\ -n)$. The l.c.m. of m and n is equal to the l.c.m. of $-m$ and n and of $-m$ and $-n$. 'Minus signs' may be ignored in calculating l.c.m.'s. L.c.m.'s are always positive.

Example. Find the l.c.m. of -8 and -12.

Solution. 24.

   1. Find the l.c.m. of 3816 and -1008.

   2. Find the l.c.m. of 18, 108 and 1008.

3. Find the g.c.d. of 111111, -12321 and 23331.

4. Find the l.c.m. of 1, 2, 3, 4, 5, 6, 7, 8, 9, 10.

5. (Harder) Prove that the l.c.m. of $n-1$, $n+1$ and n^2+1 is

$$\begin{cases} n^4 - 1 & \text{if } n \text{ is even,} \\[2mm] \dfrac{n^4 - 1}{4} & \text{if } n \text{ is odd.} \end{cases}$$

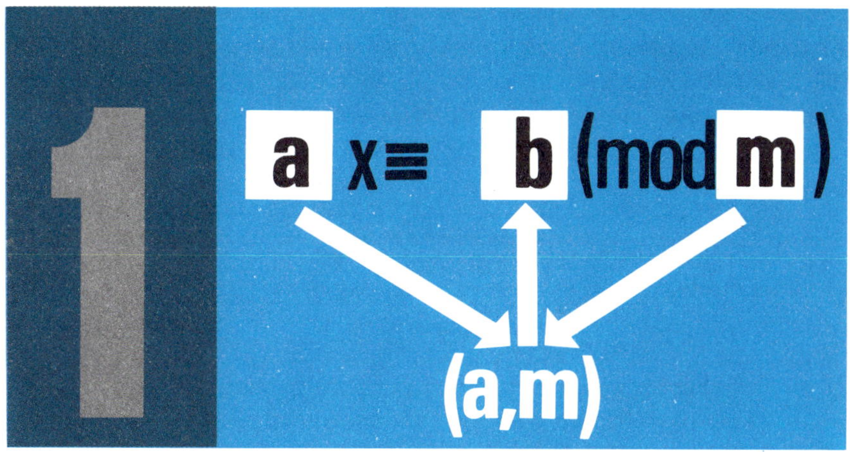

Congruence Equations of Degree 1

'I'm very sorry to hear that one of my loyal multiples feels the way 0 does. I would like to be able to allow her to divide other numbers but as the monarch I have a responsibility to the kingdom as a whole. I dread to think what the consequences would be if I allowed division by zero.

'Now royal humility prevents me from talking too much about myself so instead let me tell you about the work of my multiples. You will realise, of course, that every number is one of my multiples. One of our most important industries is that of solving congruence equations. Her Royal Highness Queen Number II will tell you about second degree equations in a moment, so let me concentrate on equations of degree 1.

'These equations are mined in puzzles and practical problems in other lands and come to us in the raw state

$$ax \equiv b \pmod{m}.$$

They first pass through a testing plant where they are subject to tests to ensure that the greatest common divisor of a and m divides b. Equations that fail to meet these rigid specifications are discarded as being insoluble. The remaining equations then go through a reverse Euclidean algorithm process in which a very powerful solvent strips the equation of its coefficient and produces the required solution.'

Method of solution

'Find a multiple of 123 which is 231 more than a multiple of 312.'

Faced with such a problem we might be tempted to use trial and error. After a few trials and as many 'errors' we would begin to realise that the trial and error method could take a long time. This is so, for the *smallest* positive solution to the problem is 5535.

Expressed algebraically, the problem is to find numbers x and y so that

$$123x = 231 + 312y$$

or equivalently,

$$123x - 231 = 312y.$$

In the last chapter we defined '$a \equiv b$ (mod m)' to mean that $a - b$ is a multiple of m. Hence we may write the above equation as

$$123x \equiv 231 \ (\text{mod } 312).$$

In this chapter we will discuss congruence equations of the form

$$ax \equiv b \ (\text{mod } m).$$

Before tackling the general case, let us consider some simple examples. What are the solutions, for example, of the equation

$$15x \equiv 8 \ (\text{mod } 21)?$$

If x is such a solution then $15x - 8 = 21y$ for some number y and so $15x - 21y = 8$. But this is impossible since $15x - 21y$ is a multiple of 3 while 8 is not. The equation therefore has no solutions.

The general case

$$ax \equiv b \ (\text{mod } m)$$

can be written as

$$ax = b + my \text{ for some } y$$

and if there are such numbers x and y, then (a, m) which divides both a and m, must divide b. Hence

$$ax \equiv b \ (\text{mod } m) \text{ has no solutions if } (a, m) \text{ does not divide } b.$$

This then gives a test for congruence equations. If an equation fails to pass the test '(a, m) divides b', we may reject it as having no solutions.

What if it passes the test? Can we be sure that there is a solution, or might there be some other condition that must be satisfied? The equation

$$93x \equiv 3 \ (\text{mod } 42),$$

for example, passes this test since (93, 42) = 3 which certainly divides
3. Now are you sure that the g.c.d. of 93 and 42 is 3? 'It is obvious by
inspection,' you say. Well, check it by Euclid's algorithm. You are sure
anyway? Never mind, check it.

$$
\begin{array}{r}
2 \\
42\overline{\smash{)}93} \\
84
\end{array}
\quad
\begin{array}{r}
4 \\
9\overline{\smash{)}42} \\
36
\end{array}
\quad
\begin{array}{r}
1 \\
6\overline{\smash{)}9} \\
6
\end{array}
\quad
\begin{array}{r}
2 \\
3\overline{\smash{)}6} \\
6 \\
\hline
0
\end{array}
$$

Hence (93, 42) = 3.

'What did I tell you?' you are probably saying; 'Of course it is 3.'
However, the real point in getting you to check it by Euclid's algorithm
was not that there was any doubt about the answer, but because the
calculations can be used to obtain a solution to the equation.

The second last division,

$$
\begin{array}{r}
1 \\
6\overline{\smash{)}9} \\
6 \\
\hline
3
\end{array}
$$

shows that $3 = 9 - 6 \times 1$. The third last division,

$$
\begin{array}{r}
4 \\
9\overline{\smash{)}42} \\
36 \\
\hline
6
\end{array}
$$

shows that $6 = 42 - 9 \times 4$. Substituting from the second of these
equations into the first, we find that

$$3 = 9 - 6 \times 1 = 9 - (42 - 9 \times 4) \times 1$$
$$= 9 \times 5 - 42.$$

The fourth last division (i.e. the first) shows that $9 = 93 - 42 \times 2$.

Hence

$$3 = 9 \times 5 - 42 = (93 - 42 \times 2) \times 5 - 42$$
$$= 93 \times 5 - 42 \times 11.$$

We have expressed the g.c.d. of 93 and 42 in the form $93x + 42y$ (with $x = 5$ and $y = -11$) by using Euclid's algorithm in reverse. In the same way it is possible to express the g.c.d. of any two numbers m and n in the form $mx + ny$.

The equation $3 = 93 \times 5 - 42 \times 11$ can now be turned into a congruence equation:

$$93 \times 5 \equiv 3 \ (\text{mod } 42).$$

Thus $x = 5$ is a solution to the equation $93x \equiv 3 \ (\text{mod } 42)$.

Example. Solve the equation $38x \equiv 6 \ (\text{mod } 106)$.

Solution

$$
\begin{array}{r}
2 \\ \hline
38\,|\,106 \\
76 \quad 1 \\ \hline
30\,|\,38 \\
30 \quad 3 \\ \hline
8\,|\,30 \\
24 \quad 1 \\ \hline
6\,|\,8 \\
6 \quad 3 \\ \hline
2\,|\,6 \\
6 \\ \hline
0
\end{array}
$$

The g.c.d. of 38 and 106 is thus 2 (as we could have seen by inspection), and

$$2 = 8 - 6$$
$$= 8 - (30 - 3 \times 8) = 8 \times 4 - 30$$
$$= (38 - 30) \times 4 - 30 = 38 \times 4 - 5 \times 30$$
$$= 38 \times 4 - 5 \times (106 - 2 \times 38)$$
$$= 38 \times 14 - 5 \times 106.$$

Hence

$$6 = 38 \times (14 \times 3) - (5 \times 3) \times 106$$
$$= 38 \times 42 - 15 \times 106$$

and so

$$38 \times 42 \equiv 6 \pmod{106}.$$

Thus $x = 42$ is a solution to the equation $38x \equiv 6 \pmod{106}$.

If (a, m) divides b then we can see how the equation $ax \equiv b \pmod{m}$ can be solved. We therefore have a test we can apply to any congruence equation which will decide whether or not a solution exists:

$$ax \equiv b \pmod{m}$$

If (a, m) divides b, there is a solution.

If (a, m) does not divide b, there are no solutions.

EXERCISE SET 4

1. Which of the following congruence equations have a solution?
   (a) $12x \equiv 34 \pmod{56}$;   (b) $34x \equiv 56 \pmod{12}$;
   (c) $65x \equiv 43 \pmod{21}$;   (d) $43x \equiv 21 \pmod{56}$.

2. Solve the congruence equation $89x \equiv 7 \pmod{65}$.

3. Solve the congruence equation $98x \equiv 7 \pmod{56}$.

4. The following problem is taken from a ninth-century Hindu manuscript:

   'Into the bright and refreshing outskirts of a forest which were full of numerous trees with their branches bent down with the weight of flowers and fruits . . . a number of travellers entered with joy. There were 63 equal heaps of fruits put together and 7 single fruits. These were divided evenly among 23 travellers. Tell me now the number of fruits in each heap.'

   Express this problem as a congruence equation and solve it.

Manipulation of congruence equations

We can often reduce the amount of computation in solving congruence equations by replacing them by equivalent but simpler equations. But to be able to manipulate congruence equations validly we must investigate the relation of congruence.

In many ways, congruence is similar to equality. The following is a list of some of the properties. These statements hold for all numbers a, b, c, d and m.

(i) $a \equiv a \pmod{m}$,
(ii) if $a \equiv b \pmod{m}$ then $b \equiv a \pmod{m}$,
(iii) if $a \equiv b \pmod{m}$ and $b \equiv c \pmod{m}$ then $a \equiv c \pmod{m}$,
(iv) if $a \equiv b \pmod{cm}$ then $a \equiv b \pmod{m}$,

(v) if $a \equiv b \pmod{m}$ and $c \equiv d \pmod{m}$ then

      (a) $a + c \equiv b + d \pmod{m}$,

      (b) $a - c \equiv b - d \pmod{m}$,

      (c) $ac \equiv bd \pmod{m}$.

The proofs of these are mostly straightforward. We can illustrate this by giving the proof of property (v). Suppose $a \equiv b \pmod{m}$ and $c \equiv d \pmod{m}$. Then $a - b = hm$ and $c - d = km$ for some numbers h and k. Now,

$$(a + c) - (b + d) = (a - b) + (c - d) = (h + k)m,$$
$$(a - c) - (b - d) = (a - b) - (c - d) = (h - k)m,$$
$$ac - bd = (a - b)c + b(c - d) = (hc + bk)m.$$

All are multiples of m and so results (a), (b) and (c) hold.

These properties show that we may multiply both sides of a congruence equation by any number. There are, however, problems in dividing through a congruence equation. For example, $2 \times 3 \equiv 1 \times 3 \pmod{3}$, but we cannot conclude that $2 \equiv 1 \pmod{3}$. It is not enough to prohibit division by zero when working with congruence equations (as it is with ordinary equations). We must only divide by numbers which are coprime with the modulus. For suppose that

$$ca \equiv cb \pmod{m}$$

where $(c, m) = 1$.

Then the equation $cx \equiv 1 \pmod{m}$ has a solution since (c, m) divides 1. As far as congruence equations modulo m are concerned, this number x is an inverse for c. We may divide by c by multiplying by x. For multiplying the first of the above two equations by x and the second by a and also by b, we obtain

$$cax \equiv cbx \pmod{m},$$
$$cax \equiv a \pmod{m},$$
$$cbx \equiv b \pmod{m},$$

and so

$$a \equiv cax \equiv cbx \equiv b \pmod{m}.$$

In many cases the process of division can be used to solve a congruence equation without using the reverse Euclidean algorithm process.

Example. Solve the equation $24x \equiv 13 \pmod{37}$.

Solution
$$24x \equiv 13 \pmod{37}$$
$$\equiv 50 \pmod{37}$$

Hence
$$12x \equiv 25 \pmod{37}$$
$$\equiv 62 \pmod{37}$$
Hence
$$6x \equiv 31 \pmod{37}$$
$$\equiv 68 \pmod{37}$$
Hence
$$3x \equiv 34 \pmod{37}$$
$$\equiv 71 \pmod{37}$$
$$\equiv 108 \pmod{37}$$
Hence
$$x \equiv 36 \pmod{37}.$$

Thus the solutions are

$$\ldots\ldots, -75, -38, -1, 36, 73, 110, \ldots\ldots$$

Example. Solve the equation $24x \equiv 14 \pmod{38}$.

Solution. Since $(24, 38) = 2$, which divides 14, the equation has solutions. We would like to divide through the equation by 2, but 2 is not coprime with the modulus. However, we may write the equation as

$$24x = 14 + 38y$$

for some number y. This is now an ordinary equation and so we may divide by 2 to obtain

$$12x = 7 + 19y.$$

Turning this back into a congruence equation we have

$$12x \equiv 7 \pmod{19}$$
$$\equiv 26 \pmod{19}$$
and so
$$6x \equiv 13 \pmod{19}$$
$$\equiv 32 \pmod{19}$$
and so
$$3x \equiv 16 \pmod{19}$$
$$\equiv 35 \pmod{19}$$
$$\equiv 54 \pmod{19}$$
and so
$$x \equiv 18 \pmod{19}.$$

The solutions are thus

$$\ldots\ldots, -39, -20, -1, 18, 37, 56, \ldots\ldots$$

The fact which is illustrated by the earlier part of the above example is that if

$$ca \equiv cb \pmod{cm}$$
then
$$a \equiv b \pmod{m}.$$

In other words we could say that a congruence equation may be divided by a divisor of the modulus, but in this case the modulus must be divided as well.

The most general result about dividing a congruence equation is that

$$ca \equiv cb \pmod{m} \text{ is equivalent to}$$

$$a \equiv b \pmod{\frac{m}{(c, m)}}.$$

We leave the proof as an exercise.

Complete solution of a congruence equation

As we saw above, congruence equations have many solutions. In discussing the reverse Euclidean algorithm process, however, we concentrated on finding just one solution. How do we go about finding them all?

Suppose that x_0 is one solution to the equation $ax \equiv b \pmod{m}$. Then $ax_0 \equiv b \pmod{m}$. For any solution, x,

$$ax \equiv b \equiv ax_0 \pmod{m}.$$

Dividing throughout by a we obtain

$$x \equiv x_0 \ [\mathrm{mod}\ m/(a, m)].$$

Conversely, for any number x which is congruent to x_0 modulo $m/(a, m)$,

$$ax \equiv ax_0 \pmod{m}$$
$$\equiv b \pmod{m}.$$

Thus having found one solution, x_0, we may write the complete solution to the equation as

$$x \equiv x_0 \ [\mathrm{mod}\ m/(a, m)].$$

Example. Find all multiples of 123 which are 231 more than a multiple of 312.

Solution. Let the multiples of 123 and 312 be $123x$ and $312y$ respectively. Then

$$123x = 231 + 312y$$

and so

$$123x \equiv 231 \pmod{312}.$$

Dividing through by 3 we obtain

$$41x \equiv 77 \pmod{104}.$$

$$
\begin{array}{r}
2 \\
41 \overline{)\,104} \\
82 \quad 1 \\
\hline
22 \overline{)\,41} \\
22 \quad 1 \\
\hline
19 \overline{)\,22} \\
19 \quad 6 \\
\hline
3 \overline{)\,19} \\
18 \\
\hline
1
\end{array}
$$

Hence
$$1 = 19 - 3 \times 6$$
$$= 19 - (22 - 19) \times 6 = 19 \times 7 - 22 \times 6$$
$$= (41 - 22) \times 7 - 22 \times 6 = 41 \times 7 - 22 \times 13$$
$$= 41 \times 7 - (104 - 2 \times 41) \times 13 = 41 \times 33 - 104 \times 13.$$

Thus
$$41 \times 33 \equiv 1 \ (\mathrm{mod}\ 104)$$
and so
$$41 \times (33 \times 77) \equiv 77 \ (\mathrm{mod}\ 104).$$

Hence $x = 33 \times 77 = 2541$ is a solution. The general solution is

$$x \equiv 2541 \ (\mathrm{mod}\ 104).$$

$$
\begin{array}{r}
24 \\
104 \overline{)\,2541} \\
208 \\
\hline
461 \\
416 \\
\hline
45
\end{array}
$$

Hence $2541 \equiv 45 \ (\mathrm{mod}\ 104)$ and so the general solution is

$$x \equiv 45 \ (\mathrm{mod}\ 104)$$
i.e.
$$x = 45 + 104n.$$

The multiple of 123 that corresponds to it is

$$123 \times 45 + 123 \times 104n$$
$$= 5535 + 12792n.$$

Hence the smallest positive answer to the problem is 5535, and the next is 18327.

1. Examine the following congruence equations, rejecting any which are insoluble. Find the complete solution to the others.
   (a) $900x \equiv 60 \pmod{1001}$, (b) $901x \equiv 61 \pmod{1001}$,
   (c) $902x \equiv 62 \pmod{1001}$, (d) $903x \equiv 63 \pmod{1001}$.

2. (a) Find the general solution to the congruence equation
   $$71x \equiv 10000 \pmod{17}.$$
   (b) Solve the following problem:

   'A cinema charges 71p admission for adults and 17p for children. At a certain session there were more adults than children and exactly £100 was taken at the box-office. How many children and how many adults were there?' [N.B. £1 = 100p]

3. (a) Find integers a and b such that
   $$79a - 37b = 1.$$
   (b) Hengest's comet returns every 79 years and Horsa's comet every 37 years. They were observed by King Arthur in A.D. 483 and A.D. 504 respectively. When will they next be visible in the same year?

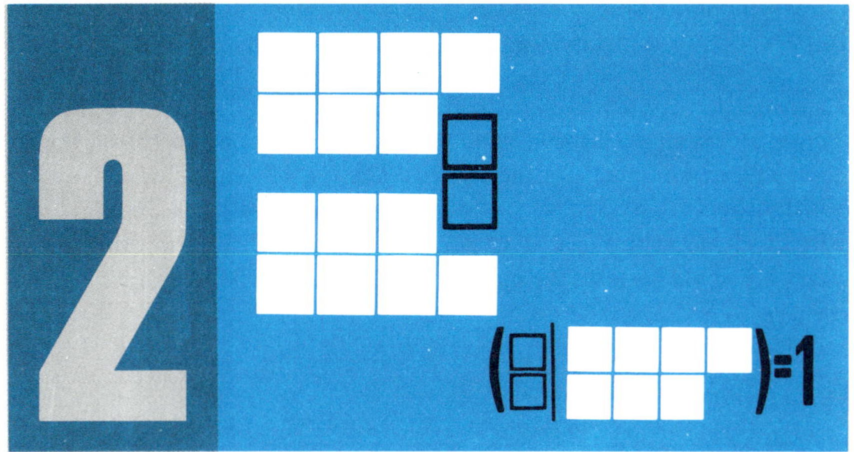

Congruence Equations of Degree 2

'My husband has asked me to talk to you about congruence equations of the second degree. I have inspected many second degree factories and talked with many of the workers and they tell me that their raw materials come in the form of equations such as

$$ax^2 + bx + c \equiv 0 \ (\mathrm{mod}\ m)$$

which grow in neighbouring lands. The stalks and foliage are separated from the roots and the square roots are then extracted. It is quite a fascinating process to watch.

'Many of these equations of degree 2 do not have square roots and so, before being processed, they are tested. I have been shown the lovely quadratic residues and I quite agree with the mathematician Gauss who once described them as the 'beautiful gems of higher arithmetic'. It is indeed wonderful to think that one can find things of such great beauty and value playing such a role in an ordinary work-a-day setting and it makes one remember that it is among the ordinary work-a-day numbers of this land of ours that our true worth lies.'

Solving the equation

'Find two consecutive odd numbers whose product is 11 more than a multiple of 111.' If the numbers are $2x + 1$ and $2x + 3$ then we may express the problem algebraically as

$$(2x + 1)(2x + 3) \equiv 11 \pmod{111}$$

i.e. $\qquad 4x^2 + 8x - 8 \equiv 0 \pmod{111}$

i.e. $\qquad x^2 + 2x - 2 \equiv 0 \pmod{111}.$

This is an example of a quadratic congruence equation. To solve such an equation, we 'complete the square'. That is, we find a perfect square which differs from $x^2 + 2x - 2$ only in the constant term. Such a perfect square in this case is $x^2 + 2x + 1 = (x + 1)^2$. We may therefore write the above congruence equation as

$$(x + 1)^2 - 3 \equiv 0 \pmod{111}$$

i.e. $\qquad (x + 1)^2 \equiv 3 \pmod{111}.$

If we put $y = x + 1$, our problem reduces to that of solving the congruence equation

$$y^2 \equiv 3 \pmod{111}$$

or, in other words, of finding the square roots of 3 modulo 111. Now there is no whole number whose square is 3. But there are numbers whose square is congruent to 3 modulo 111. We defer discussion on the way we find square roots for a given modulus and simply state a fact, which may be easily verified: $15^2 \equiv 3 \pmod{111}$. Hence $y \equiv 15 \pmod{111}$ is a solution to the above congruence equation and so, $x \equiv 14 \pmod{111}$ is a solution to the original equation. In other words, any number of the form $14 + 111n$ is a solution. This gives rise to the consecutive odd numbers $222n + 29$, $222n + 31$. For example 29 and 31 are consecutive odd numbers whose product, 899, is 11 more than a multiple of 111.

There are solutions other than $222n + 29$, $222n + 31$, since 15 is not the only square root of 3 modulo 111. Another square root is -15. That is, $y \equiv -15 \pmod{111} \equiv 96 \pmod{111}$ is another possibility. This corresponds to $x \equiv 95 \pmod{111}$ and gives rise to consecutive odd numbers of the form $222n + 191$, $222n + 193$. Hence 191, 193 etc. are also possible solutions to the problem.

Consider the general equation

$$ax^2 + bx + c \equiv 0 \pmod{m}.$$

We shall only discuss the case where $(2a, m) = 1$. Multiplying the equation by $4a$ we obtain

$$4a^2 x^2 + 4abx + 4ac \equiv 0 \pmod{m},$$

and completing the square, we can write this as

$$(2ax + b)^2 + 4ac - b^2 \equiv 0 \ (\text{mod } m),$$

i.e. $\qquad y^2 \equiv b^2 - 4ac \ (\text{mod } m)$

where $\qquad y \equiv 2ax + b \ (\text{mod } m)$

i.e. $\qquad 2ax \equiv y - b \ (\text{mod } m).$

Conversely, we may show that for any x satisfying these equations, $4a(ax^2 + bx + c) \equiv 0 \ (\text{mod } m)$ and since $(2a, m) = 1$, $(4a, m) = 1$ and so we may divide by $4a$ to obtain $ax^2 + bx + c \equiv 0 \ (\text{mod } m)$. The complete solutions to this equation is therefore given by the complete solutions to

$$2ax \equiv y - b \ (\text{mod } m)$$

for all y satisfying the equation

$$y^2 \equiv b^2 - 4ac \ (\text{mod } m).$$

For every square root of $b^2 - 4ac$, modulo m, we may substitute into the first of these equations and then, since $(2a, m) = 1$, we may solve this equation using the methods of the last chapter.

Example. Find two consecutive numbers, other than 3 and 4, whose product is 12 more than a multiple of 111.

Solution. If the numbers are x and $x + 1$, the equation to be solved is

$$x(x + 1) \equiv 12 \ (\text{mod } 111)$$

i.e. $\qquad x^2 + x - 12 \equiv 0 \ (\text{mod } 111).$

The complete solution is given by

$$2x \equiv y - 1 \ (\text{mod } 111)$$

for all y satisfying the equation

$$y^2 \equiv 1 + 48 \ (\text{mod } 111)$$
$$\equiv 49 \ (\text{mod } 111).$$

Solutions to this equation are $y = \pm 7$. Taking $y = 7$, we have the equation

$$2x \equiv 7 - 1 \ (\text{mod } 111) \equiv 6 \ (\text{mod } 111)$$

i.e. $\qquad x \equiv 3 \ (\text{mod } 111).$

The consecutive numbers that correspond to this equation are $111n + 3$ and $111n + 4$. For $y = -7$ we get the equation

$$2x \equiv -8 \ (\text{mod } 111)$$

i.e. $\qquad x \equiv -4 \ (\text{mod } 111).$

The consecutive numbers that correspond to this are $111n - 4$,

$111n - 3$. Hence consecutive numbers of the form $111n + 3$, $111n + 4$ and of the form $111n - 4$, $111n - 3$ are solutions. The smallest positive solutions (other than 3, 4) are 97, 98.

1. Solve the following congruence equations:
   (a) $x^2 + 5x - 3 \equiv 0 \pmod{37}$; (b) $2x^2 + 7x + 5 \equiv 0 \pmod{37}$;
   (c) $5x^2 + 8x + 3 \equiv 0 \pmod{37}$; (d) $15x^2 + 7x \equiv 0 \pmod{37}$.

2. Solve the congruence equation
$$x^2 + x + 1 \equiv 0 \pmod{37}.$$
   N.B. The square roots of 34 modulo 37 are ± 16.

3. Find four consecutive numbers such that the product of the first two plus the product of the second two is a multiple of 37.

Finding square roots

The following examples illustrate methods for finding square roots for a given modulus.

Example. Find the square roots of 33 modulo 97.

Solution. $33 \equiv 130 \equiv 227 \equiv 324 = 18^2$, so ± 18 (i.e. 18, 79) are square roots of 33 modulo 97. (The congruences are to the modulus 97.)

Example. Find the square roots of 24 modulo 101.

Solution. $24 \equiv 125 \equiv 226 \equiv 327 \equiv 428 \equiv 529 = 23^2$, so ± 23 (i.e. 23, 78) are square roots of 24 modulo 101.

The method is simply to add the modulus repeatedly to the number until a perfect square is obtained. A list of perfect square is given in table 1.

Table 1 : Table of Squares from 100 to 2401

	+0	+1	+2	+3	+4	+5	+6	+7	+8	+9
10	100	121	144	169	196	225	256	289	324	361
20	400	441	484	529	576	625	676	729	784	841
30	900	961	1024	1089	1156	1225	1296	1369	1444	1521
40	1600	1681	1764	1849	1936	2025	2116	2209	2304	2401

Example. Find the square roots of 16 modulo 111.

Solution. One easily obtains ± 4, i.e. 4 and 107, as square roots. However, there are other square roots in this case. For, modulo 111,

$$16 \equiv 127 \equiv 238 \equiv 349 \equiv 460 \equiv 571 \equiv 682 \equiv 793 \equiv 904 \equiv$$
$$1015 \equiv 1126 \equiv 1231 \equiv 1348 \equiv 1459 \equiv 1570 \equiv 1681 = 41^2.$$

Hence ± 41 (i.e. 41 and 70) are also square roots of 16 modulo 111.

In this respect congruence arithmetic differs from ordinary arithmetic. In congruence arithmetic, a number may have more than two square roots (none of which are congruent to one another) while in ordinary arithmetic numbers can have only two square roots.

EXERCISE SET 7

1. Find square roots of the following numbers modulo 101:
   (a) 14; (b) 23; (c) 71; (d) 82; (e) 95.

2. Solve the congruence equation
$$11x^2 + 7x + 2 \equiv 0 \ (\text{mod } 41).$$

3. Find all of the square roots of 3 modulo 41.

Quadratic residues

Not every number has a square root to a given modulus. For example, there is no number x such that

$$x^2 \equiv 2 \ (\text{mod } 3).$$

This is because
$$0^2 \equiv 0 \ (\text{mod } 3),$$
$$1^2 \equiv 1 \ (\text{mod } 3),$$
and
$$2^2 \equiv 1 \ (\text{mod } 3).$$

Other numbers are congruent to 0, 1 or 2 modulo 3 and so their squares will be congruent to 0 or 1, modulo 3. The only numbers between 0 and 3, inclusive, that have square roots modulo 4 are 0 and 1 since

$$0^2 \equiv 2^2 \equiv 0 \ (\text{mod } 4)$$
and
$$1^2 \equiv 3^2 \equiv 1 \ (\text{mod } 4).$$

So, modulo 3 and 4, 2 has no square roots. However, modulo 7, it has since $3^2 \equiv 2 \ (\text{mod } 7)$.

If a number a has a square root modulo m we say that a is a *quadratic residue* modulo m. Thus the quadratic residues modulo 7 are 0, 1, 2, 4 (and any numbers which are congruent to them modulo 7).

The numbers 3, 5 and 6 are not quadratic residues.

We define the symbol $(a|m)$ to have the value 0 or 1 according to whether a is a quadratic residue modulo m, or not.

Examples

$$(0|7) = 0, (1|7) = 0, (2|7) = 0, (3|7) = 1,$$
$$(4|7) = 0, (5|7) = 1, (6|7) = 1.$$

To determine whether or not a number a is a quadratic residue modulo m we need only calculate $(a|m)$. This we can do using a number of facts about quadratic residues. These were first stated by Legendre, but property 7 was first proved by Gauss at the end of the eighteenth century.

Quadratic Residues

$$(a|m) = \begin{cases} 0 \text{ if } x^2 \equiv a \ (\text{mod } m) \text{ has a solution.} \\ 1 \text{ if } x^2 \equiv a \ (\text{mod } m) \text{ has no solution.} \end{cases}$$

1. If $a \equiv b \ (\text{mod } m)$ then $(a|m) = (b|m)$.

2. If $(k, m) = 1$, then $(k^2 a|m) = (a|m)$.

3. If m is odd then $(a|2^r m) = 0$ if and only if $(a|p) = 0$ for all prime divisors, p, of m and $(a|2^r) = 0$.

4. If $r \geqslant 3$, and $0 < a < 2^r$ then $(a|2^r) = 0$ if and only if a can be expressed in the form $4^s t$ where $t \equiv 1 \ (\text{mod } 8)$.

5. If $r \leqslant 2$, $(a|2^r) = 0$ if and only if $a \equiv 0 \ (\text{mod } 2^r)$ or $a \equiv 1 \ (\text{mod } 2^r)$.

6. If p is an odd prime,
$$(ab|p) \equiv (a|p) + (b|p) \ (\text{mod } 2).$$

7. If p and q are odd primes,
$$(q|p) \equiv (p|q) + (-1|p)\,(-1|q) \ (\text{mod } 2).$$

8. If p is an odd prime,
$$(-1|p) \equiv \frac{p-1}{2} \ (\text{mod } 2).$$

9. If p is an odd prime,
$$(2|p) \equiv \frac{1}{2}\left(\frac{p-1}{2}\right)\left(\frac{p+1}{2}\right) \ (\text{mod } 2).$$

10. $(0|m) = (1|m) = 0$ for all m.

Some of these facts, such as numbers 1, 2, 5, 10 are easily proved. For example, since $0^2 \equiv 0 \ (\text{mod } m)$ and $1^2 \equiv 1 \ (\text{mod } m)$ for any modulus m, $(0|m)$ and $(1|m)$ are both 0. Others are more difficult to prove. Gauss said of property 7, which is known as the 'quadratic reciprocity law': 'It tortured me for a whole year and eluded the most strenuous efforts before, finally, I got the proof . . . '. He referred to this law as 'theorem aureum', the golden theorem, and the 'gem of higher arithmetic'.

The ten properties of quadratic residues are sufficient to enable $(a|m)$ to be calculated for any numbers a and m. The procedure is as follows. Property 1 can be used to ensure that a is non-negative and less than m and property 2 can be used to remove any square factors in a. The next step is to factorise the modulus. Property 3 enables us to reduce the problem to the case where the modulus is an odd prime or a power of 2. Properties 4 and 5 deal with the case where the modulus is a power of 2. The next step is to factorise a into prime factors and use property 6 to reduce the problem to the case $(q|p)$ where p and q are primes and where p is odd. Property 9 takes care of the case where $q = 2$. If q is odd and greater than or equal to p, rule 1 may be used to replace it by something smaller than p. If q is odd and less than p, property 7 enables us to express $(q|p)$ in terms of $(p|q)$, $(-1|p)$ and $(-1|q)$. Property 8 can be used to calculate $(-1|p)$ and $(-1|q)$ and, p being greater than q, can be replaced in (p, q) by something smaller than q, by property 1. The process continues in this way.

The following examples illustrate the process.

Example. Does the equation $x^2 \equiv 476 \pmod{400}$ have a solution?

Solution. $(476|400) = (76|400)$ by property 1,
$$= (76|16 \times 25)$$
$$= 0 \text{ if and only if } (76|5) \text{ and } (76|16) \text{ are}$$
$$\text{both } 0 \text{ (property 3).}$$

Now $(76|5) = (1|5)$ by property 1,
$$= 0 \text{ by property 10.}$$
$(76|16) = (12|16)$ by property 1,
$$= (4 \times 3|16) = 1 \text{ by property 4, since } 3 \not\equiv 1 \pmod{8}.$$

Hence $(476|400)$ is not zero. It therefore must be 1 and so the equation $x^2 \equiv 476 \pmod{400}$ has no solution.

Example. Is 33 a quadratic residue modulo 41?

Solution. (All congruences here are to the modulus 2.) Since 41 is an odd prime, it follows from property 6 that

$$(33|41) \equiv (3|41) + (11|41).$$

By property 7,

$$(3|41) \equiv (41|3) + (-1|3)\,(-1|41)$$
$$\equiv (2|3) + 1 \times 20 \text{ using property 8}$$
$$\equiv (2|3)$$
$$\equiv (-1|3) \text{ by property 1,}$$
$$\equiv 1 \text{ by property 8.}$$

$$(11|41) \equiv (41|11) + (-1|11)\,(-1|41)$$
$$\equiv (8|11) + 5 \times 20$$
$$\equiv (8|11)$$
$$= (2|11) \text{ by property 2,}$$
$$\equiv 5 \times 3 \text{ by property 9,}$$
$$\equiv 1.$$

Thus $(33|41) \equiv (3|41) + (11|41) \equiv 1 + 1 \equiv 0$, and so 33 is a quadratic residue modulo 41.

Table 2 can be used to assist in these calculations. It lists the first 50 primes. Those primes p for which $(-1|p) = 0$ are printed in blue and those for which $(-1|p) = 1$ are printed in black. The quadratic reciprocity law (property 7) can be interpreted as follows: 'Provided at least one of the primes p and q is in blue, $(p|q) = (q|p)$. Otherwise they are different.'

$(-1|p) = 0$ if p is in blue

$\qquad\quad$ 1 if p is in black

$(p|q)\quad = (q|p)$ if and only if at least one of p, q is in blue.

2	3	5	7	11	13	17	19	23	29
31	37	41	43	47	53	59	61	67	71
73	79	83	89	97	101	103	107	109	113
127	131	137	139	149	151	157	163	167	173
179	181	191	193	197	199	211	223	227	229

Example. Find $(151|227)$.

Solution. $(151|227) \neq (227|151)$ since neither prime is in blue in table 2. $(227|151) = (76|151) = (19|151)$, by property 2. Now $(19|151) \neq (151|19)$ since 19 and 151 are both in black. $(151|19) = (18|19) = (-1|19) = 1$ since 19 is in black. Thus $(19|151) = 0$ and so $(227|151) = 0$. Hence $(151|227) = 1$.

Example. Find $(139|1145)$.

Solution. Since $1145 = 5 \times 229$, $(139|1145) = 0$ if and only if $(139|5) = 0$ and $(139|229) = 0$. Now $(139|5) = (4|5) = 0$.

$$
\begin{aligned}
(139|229) &= (229|139) \text{ since 229 is in blue,} \\
&= (90|139) \\
&= (10|139) \text{ on taking out the square factor, 9,} \\
&\equiv (2|139) + (5|139) \\
&\equiv 69 \times 35 + (139|5) \text{ by properties 9 and 7,} \\
&\equiv 1 + (4|5) \\
&\equiv 1 + 0 \\
&\equiv 1.
\end{aligned}
$$

Hence $(139|1145)$ is not 0. It is therefore equal to 1.

EXERCISE SET 8

1. Find the following:
   (a) $(79|109)$;   (b) $(95|139)$;   (c) $(89|398)$;
   (d) $(822|350)$;   (e) $(73|128)$.

2. Find the following:
   (a) $(604|47 \times 47)$;   (b) $(37 \times 103 \mid 29 \times 229)$;
   (c) $(167 \times 167 \times 173 \mid 103 \times 211 \times 227)$.

3. Has the equation $7x^2 + 5x - 10 \equiv 0 \,(\text{mod } 29)$ solutions?

4. Prove that if $p = 4k + 1$ is prime then
$$(2|p) \equiv k \,(\text{mod } 2).$$

5. (Harder) Prove that if p and q are odd primes such that $q = 2p + 1$, then $(p|q) = (-1|p)$.

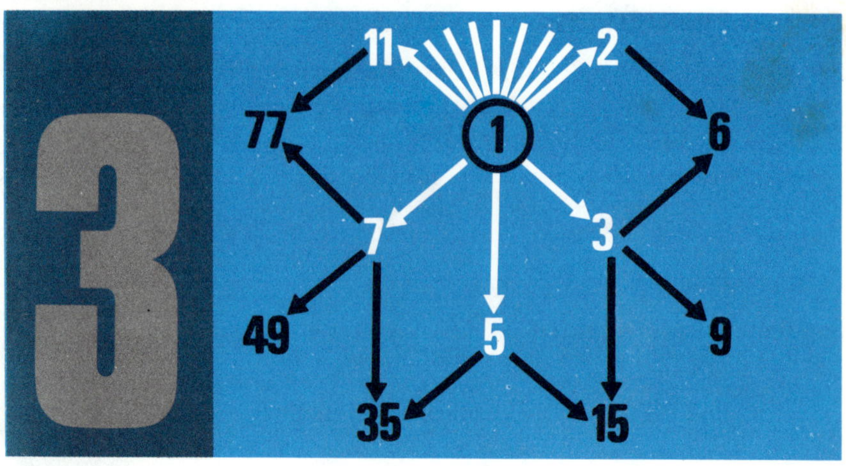

Prime Numbers

'Hello there. How are you all? I'm prime. Ha! Ha! In the prime of life one might say. Now you may ask what's so important about being prime. Well let me tell you that we prime numbers, along with King Number I, are the parents of the number nation. Why if some great calamity were to strike today wiping out all other numbers, we could replace them by multiplying among ourselves. Oh, we couldn't replace Dunce Nought but then nobody would miss her anyway.

'Well that's a quick run down on primes in general. Now let me tell you about two groups of primes with which I am very much involved. Firstly, I am president of the Mersenne Society — a *very* exclusive club. To be eligible for membership of this society, a prime must be of the form $2^n - 1$. Our membership at the moment stands at only a little over 20. We are dedicated to bringing perfect numbers into the world. I am, myself, the father of '6', the world's first and most beautiful perfect number.

'I am also the managing director of a small but progressive firm, The Constructible Polygon Company, which makes regular polygons with ruler and compass. I could go on for hours telling you about the wonderful work we do, but if I did my assistant manager, number 5, would have nothing to talk about so I'll leave it to him to tell you.'

Factorisation into primes

The sequence of primes,* 2, 3, 5, 7, 11, 13, 17, 19, 23 etc. has fascinated mathematicians throughout the ages. These numbers are the atoms of the number universe, or to use another analogy, the ancestors of the number nation. Every positive number, n, (apart from 0 or 1) can be uniquely factorised into primes. (Here we include 'factorisations' involving just one factor. Thus '2 = 2' is the factorisation of 2 into primes.) In other words, if $n > 1$, we can write n in the form

$$n = p_1 p_2 \ldots p_r$$

where $p_1, p_2 \ldots \ldots, p_r$ are primes. Moreover, apart from a rearrangement of factors, there is only one possible factorisation. There is, for example, no other way of factorising 600 into primes than

$$600 = 2 \times 2 \times 2 \times 3 \times 5 \times 5.$$

If equal factors are collected together, a number n which is greater than 1 can be written in the form

$$n = p_1^{a_1} p_2^{a_2} \ldots \ldots p_r^{a_r}$$

where $p_1, p_2, \ldots, p_r$ are different primes. Again, apart from a rearrangement of factors, there is only one way of writing n in this way. For example, 600 is expressed in this form as

$$600 = 2^3 \times 3 \times 5^2.$$

How many primes?

It has been known for a long time that the number of primes is infinite. The following proof is attributed to Euclid and is often held up as a superb example of conciseness and elegance in a mathematical proof.

Suppose that there is a largest prime, p. Then the number

$$N = 2 \times 3 \times 5 \times 7 \times 11 \times \ldots \times p + 1$$

is not divisible by any prime since it leaves a remainder of 1 when divided by any prime. But every number which is bigger than 1 is divisible by some prime. We therefore have a contradiction and so hence there could not be a largest prime. There must therefore be infinitely many primes.

Although the list of primes never terminates, primes gradually become scarcer and scarcer as we search for them among larger and larger numbers. For example, between 1 and 1000 there are 168 primes, between

*The numbers $-2, -3, -5, \ldots$ are also prime numbers but throughout the remainder of this book we shall use 'prime' to mean a *positive* prime number.

$10^6 + 1$ and $10^6 + 1000$ (inclusive) there are 75 primes and between $10^{12} + 1$ and $10^{12} + 1000$ (inclusive) there are only 37 primes.

Sieving out the primes

Table 2 on page 34 lists the primes up to 229. Tables of primes up to 10 million have been compiled. Such tables can be prepared using a method known as the 'Sieve of Eratosthenes'.

To find the primes in the first n numbers, write down the first n numbers in order. Cross out 1, since 1 is not prime. Put a circle around 2, since 2 is prime. Now go through the list striking out every second number, starting with 4. These clearly are not primes since they are divisible by 2. Go back to the beginning and circle the first number which has not been circled or crossed out. This, of course, will be the number 3. Now cross out every 3rd number, starting with 6. Go back to the beginning and circle the first number which has not been circled or crossed out. This will be 5. Now cross out every 5th number etc. Continue this process until you have found a prime which exceeds $\sqrt{n}$. All of the numbers that are left at this stage may then be circled for they must all be prime. This is because *every number $m > 1$ which is not prime, has a prime divisor which is less than or equal to $\sqrt{m}$.*

If a number m is not crossed out at this last stage, it has no prime divisors less than or equal to $\sqrt{n}$. Since $m \leqslant n$, $\sqrt{m} \leqslant \sqrt{n}$, so such a number m certainly has no prime divisors less than $\sqrt{m}$. By the above theorem, such a number must be prime.

Why must every number, m, which is bigger than 1, be either prime or have a prime divisor which is less than or equal to $\sqrt{m}$? Well, if m were not prime we could write

$$m = ab$$

where a and b are each bigger than 1. If both a and b were bigger than $\sqrt{m}$, then ab would be bigger than $(\sqrt{m})^2 = m$, a contradiction. Hence at least one of a and b must be less than or equal to $\sqrt{m}$. Suppose for example that $a \leqslant \sqrt{m}$. Since a is bigger than 1, it is divisible by a prime number, say p. Then p divides m and $p \leqslant a \leqslant \sqrt{m}$.

Example. Carry out the 'Sieve of Eratosthenes' process to find the primes among the first 25 numbers.

Solution. The successive stages in the process are as follows:

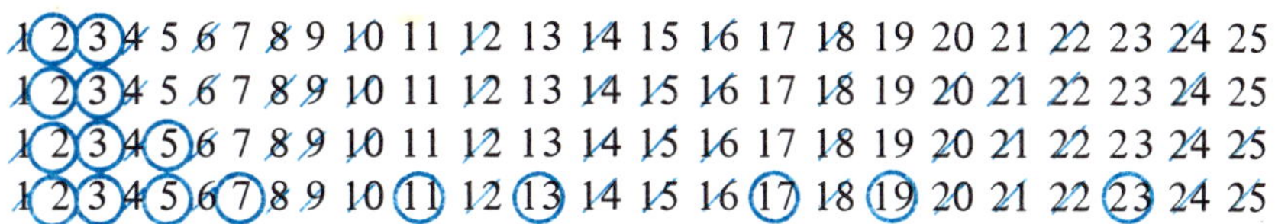

Formulae for primes

To find the 1000th prime by this method would involve finding all of the primes up to the 1000th and would clearly be a lengthy process. It would be nice to have a formula for the n'th prime so that when we substituted $n = 1000$ we could easily compute the 1000th prime. However, despite great effort throughout the centuries to find such a formula, none has ever been found.

A less ambitious request would be for a formula that produces not necessarily all of the primes, but *only* primes, i.e. a formula in terms of n which is prime for all values of n. Such a formula is

$$4 + (-1)^n.$$

This is not a very interesting formula, however, because it gives only two primes: 3 and 5. What we seek is a formula that takes infinitely many different values, all of which are prime. Unfortunately no such formula is known.

One formula which comes close to this goal is

$$n^2 + n + 41.$$

For values of n from 0 to 39, it is prime. However, for $n = 40$, it becomes 41×41 and for $n = 41$ it is 41×43. For larger values of n it is sometimes prime and sometimes *composite* (numbers bigger than 1 which are not prime are called composite numbers).

Even when $n^2 + n + 41$ is not prime it has no small prime divisors. The smallest prime divisor is always at least 41. For suppose that the prime, p, divides $n^2 + n + 41$ for some n. Then the congruence equation

$$n^2 + n + 41 \equiv 0 \pmod{p}$$

has a solution.

The solutions would be of the form

$$2n \equiv y - 1 \pmod{p}$$

where

$$y^2 \equiv -163 \pmod{p}.$$

Thus $(-163|p) = 0$. A check for prime values of p up to 37 inclusive shows that $(-163|p) = 1$ for such values. Hence $p \geqslant 41$.

The great mathematician Fermat thought he might have had a formula which produces only primes:

$$F_n = 2^{2^n} + 1.$$

For $n = 0, 1, 2, 3$ and 4, $F_n = 3, 5, 17, 257$ and 65537 respectively and all of these are prime. Fermat conjectured that F_n might be always prime. However, had he calculated and tested F_5 he would have discovered that it has 641 as a factor. So F_5 is composite.

What is surprising is that no other values of n that have been tested yield primes. F_n has been examined for $n = 0, 1, 2, 3, \ldots, 13$ and for several larger values and except for the first five values, they have all been composite.

The reason why only relatively few values of n have been tested is that F_n increases exceedingly rapidly with n. For example F_{13} has over 2500 digits.

Primes of the form $2^m + 1$ are called *Fermat primes*. We can show that in fact all Fermat primes have the form

$$2^{2^n} + 1.$$

For suppose that $2^m + 1$ is a Fermat prime and suppose that m is not a power of 2. Let p be an odd prime divisor of m. Then $m = pr$ for some number r. Algebraically we can factorise $x^p + 1$ as follows:

$$x^p + 1 = (x + 1)(x^{p-1} - x^{p-2} + \ldots + x^2 - x + 1).$$

Substituting $x = 2^r$ we get a factorisation for $2^m + 1$, contradicting the assumption that $2^m + 1$ is prime. Hence m must be a power of 2 and so the Fermat prime has the form $2^{2^n} + 1$. Thus the only known Fermat primes are $3, 5, 17, 257,$ and 65537. We shall discuss Fermat primes further in chapter 5.

Mersenne primes

Primes of the form $2^n - 1$ are the companions to the Fermat primes. They are called Mersenne primes and there are rather more of them known. To date about 25 Mersenne primes have been discovered. The first eight of them are

$$3, 7, 31, 127, 8191, 131071, 524287, 2147483647.$$

Mersenne primes play an important role in the theory of *perfect* numbers. These are numbers which are equal to the sum of their *proper* divisors — that is, the sum of all their divisors excluding themselves.

The number 6 is a perfect number since

$$6 = 1 + 2 + 3.$$

It is the first perfect number. The next is 28. It is perfect since

$$28 = 1 + 2 + 4 + 7 + 14.$$

Every perfect number has the form

$$\frac{p(p + 1)}{2}$$

where p is a Mersenne prime. The perfect numbers 6 and 28, for example, can be expressed in this form:

$$6 = \frac{3 \times 4}{2}$$

$$28 = \frac{7 \times 8}{2}.$$

We shall discuss perfect numbers further in chapter 6.

EXERCISE SET 9

1. Use the 'Sieve of Eratosthenes' to find the primes up to 200.

2. Prove that all composite numbers of the form
$$n^2 + 1$$
are divisible by some prime less than n.
[HINT: $n^2 + 1 < (n + 1)^2$ if $n \geqslant 1$.]

3. We call a number, k, *primitive* if $k > 3$ and
$$n^2 + n + k$$
is prime for $n = 0, 1, 2 \ldots k - 2$. Prove that if k is primitive then k is prime and $k \equiv 2 \pmod 3$.
[HINT: Examine the cases $n = 0$ and $n = 1$.]

4. Find all of the primitive numbers less than 100.
[HINT: Use question 3.]

   NOTE: It is not known whether or not there are any other primitive numbers. It is claimed that if there is another, there is at most one and it is bigger than 1250000000.

5. (Harder) Prove that for all number $n \geqslant 3$, there is at least one prime between n and $n!$ where $n!$ denotes $1 \times 2 \times 3 \times \ldots \ldots \times n$.

[HINT: Examine the prime divisors of $n! - 1$.]

NOTE: It has been proved, in fact, that for all numbers $n \geqslant 2$ there is at least one prime between n and $2n$.

6. Prove that if $2^{2n} - 1$ is a prime (and so a Mersenne prime) then $n = 1$. [HINT: Prove that if $n > 1$ then $2^{2n} - 1$ is not prime.]

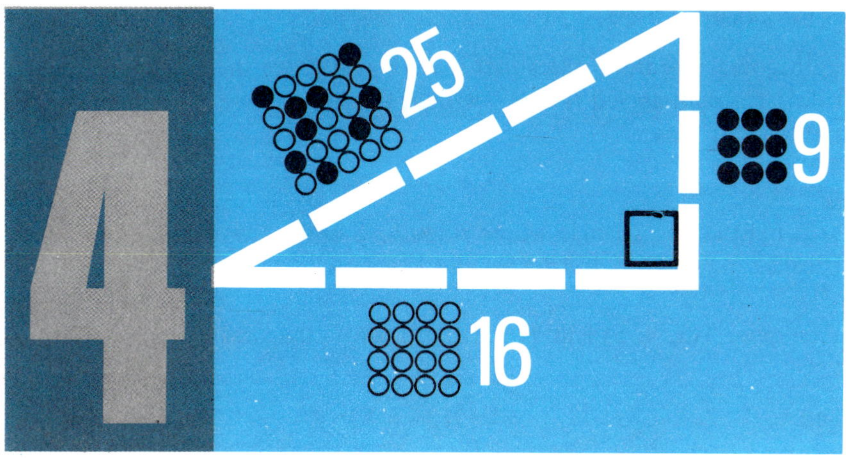

Squares

'Hullo all. My number is 4 and that makes me a perfect square don't it? I s'pose you're all looking at me and wondering what the old girl does for a crust. Well you might say I earns my crust by helping others to cut theirs. You see I makes knives — kitchen knives only, mind, none of your cloak and dagger stuff.

'The design we use for blades is a right-angled triangle with the cutting edge along the hippopotenuse, or whatever they call it. This is where we squares come in. For somebody once said — I can't pronounce his name so I won't — he said, the square on the hippopotenuse is equal to the sum of the squares on the other two sides.

'Now I 'xpect you've all heard of the 3, 4, 5 blade, but this is too short for most jobs. The 5, 12, 13 or the 7, 24, 25 are much better.

'I'll never forget the day when my good friend, 16, dropped in with the biggest Chinese Fortune Cookie you've ever seen. 'Can you give me a knife that'll cut this open?' she says. 'Hang on,' I says, 'I'll try.' I has a go with my sharpest knife but couldn't seem to make much of an impression.

'Where did you get hold of this?' I says. '4', she says, I'll tell you. This fellow comes to my door and says there's a fortune inside this cookie, and the money's mine if I can open it. I tried and couldn't cut it open. He said if I gave him my money he'd bake it in a fortune cookie for me to keep it safe. I asked him for the recipe and what do you think he did? He asked to see my cook book and when he saw it he said he couldn't write it out cause the margin wasn't big enough.'

Pythagorean triangles

Pythagoras' Theorem states that if X, Y and Z are the lengths of the sides of a right-angled triangle, with Z the hypotenuse (i.e. opposite the right angle) then

$$X^2 + Y^2 = Z^2.$$

A right-angled triangle whose sides X, Y and Z are whole numbers is called a *Pythagorean triangle* and is denoted by T(X, Y, Z).

Examples. T(3, 4, 5) and T(5, 12, 13) are Pythagorean triangles since

$$3^2 + 4^2 = 5^2$$

and

$$5^2 + 12^2 = 13^2.$$

The general formula for producing Pythagorean triangles is

$$\left. \begin{array}{l} X = k(m^2 - n^2) \\ Y = 2kmn \\ Z = k(m^2 + n^2). \end{array} \right\} \quad \text{where } m > n > 0,\ k > 0. \qquad (1)$$

If X, Y and Z have this form then

$$\begin{aligned} X^2 + Y^2 &= k^2(m^2 - n^2)^2 + 4k^2 m^2 n^2 \\ &= k^2(m^4 + n^4 - 2m^2 n^2 + 4m^2 n^2) \\ &= k^2(m^4 + n^4 + 2m^2 n^2) \\ &= k^2(m^2 + n^2)^2 = Z^2. \end{aligned}$$

Hence if X, Y and Z have the above form, the triangle $T(X, Y, Z)$ is Pythagorean.

Examples. The following triangles are Pythagorean:

$$\begin{array}{ll} T(21, 20, 29) & (k = 1, m = 5, n = 2) \\ T(16, 30, 34) & (k = 1, m = 5, n = 3) \\ T(9, 12, 15) & (k = 3, m = 2, n = 1) \end{array}$$

It is possible for different values of k, m and n to give the same Pythagorean triangle, possibly with x and y interchanged, as the following example shows.

Example. The following combinations of values of k, m and n all give the Pythagorean triangle $T(12, 16, 20)$:

$$\begin{array}{l} k = 1, m = 4, n = 2; \\ k = 2, m = 3, n = 1; \\ k = 4, m = 2, n = 1. \end{array}$$

If the sides of a Pythagorean triangle are coprime with one another, we say that the triangle is *primitive*. For a primitive Pythagorean triangle, $k = 1$ and $(m, n) = 1$. Moreover one of m, n is odd and the other is even, for if they were both odd or both even, x, y, and z would all be even and so not coprime.

Example. $T(5, 12, 13)$ is primitive, $T(9, 12, 15)$ is not.

No other pythagorean triangles

We now show that the formula (1) gives all Pythagorean triangles. Suppose that

$$X^2 + Y^2 = Z^2 \tag{2}$$

We shall show that numbers k, m and n can be found so that X, Y and Z are given by the formula (1).

Reduction to primitive case

Let $k = (X, Y)$. Then k divides both X and Y and so k^2 divides Z^2 whence k divides Z. Put

$$\left.\begin{array}{l} X = kx \\ Y = ky \\ Z = kz \end{array}\right\} \tag{3}$$

Then, dividing through the equation (2) by k^2, we obtain

$$x^2 + y^2 = z^2 \tag{4}$$

Moreover, since x and y are what are left of X and Y when their g.c.d. is removed, x and y are coprime.

z is odd

Now if both x and y are odd then

$$x^2 \equiv y^2 \equiv 1 \ (\text{mod } 4)$$

and so
$$z^2 \equiv 2 \ (\text{mod } 4),$$

which is impossible. (If z is even, $z^2 \equiv 0 \ (\text{mod } 4)$ and if z is odd, $z^2 \equiv 1 \ (\text{mod } 4)$.) So x and y are not both odd. Furthermore, they are not both even since $(x, y) = 1$. Hence one of x and y is odd, say x, and the other is even, say y. Hence $x^2 + y^2$ is odd and so z is odd.

Definition of a, b and c

Since y is even we may put

$$y = 2a \qquad (5)$$

for some number a, and substituting in (4) we obtain

$$x^2 + 4a^2 = z^2$$

and so

$$4a^2 = z^2 - x^2 = (z - x)(z + x). \qquad (6)$$

Now z and x are both odd so $z - x$ and $z + x$ are even. Let

$$\left. \begin{array}{l} z - x = 2b \\ z + x = 2c \end{array} \right\} \qquad (7)$$

Substituting in (6) we get $4a^2 = 4bc$, whence

$$a^2 = bc \qquad (8)$$

Now from (7), $\qquad\qquad 2c - 2b = 2x$

and so

$$x = c - b. \qquad (9)$$

b and c are coprime

Let $d = (b, c)$. Then d divides both b and c and so divides x, by (9). Also, d divides a by (8) and so divides y by (5). Hence d is a common divisor of x and y. But $(x, y) = 1$. Hence $d = 1$ and so b and c are coprime.

b and c are perfect squares

Let p be any prime dividing b and let p^r be the largest power of p dividing a. Then p^{2r} will be the largest power of p dividing a^2. Since $(b, c) = 1$, p does not divide c and so by (8), p^{2r} is the largest power of p dividing b. Since this holds for all prime divisors of b, b is a perfect square. Similarly c is also a perfect square.

Put

$$\left. \begin{array}{l} b = n^2 \\ c = m^2 \end{array} \right\} \qquad (10)$$

Reassembling

By (9), $\qquad\qquad x = m^2 - n^2.$

By (8), $\qquad\qquad a^2 = m^2 n^2$

whence $a = mn$ and so, by (4),

$$y = 2mn.$$

From (7), $$2z = 2b + 2c = 2(m^2 + n^2)$$

and so $$z = m^2 + n^2.$$

Finally, substituting into (3),

$$X = k(m^2 - n^2), \quad Y = 2kmn, \quad Z = k(m^2 + n^2).$$

EXERCISE SET 10

1. Find the Pythagorean triangles given by the following sets of values of k, m and n:
   (a) $k = 1$, $m = 7$, $n = 4$; (b) $k = 1$, $m = 5$, $n = 3$;
   (c) $k = 2$, $m = 3$, $n = 2$.

2. Prove that the perimeter of a Pythagorean triangle has the form
$$2km(m + n).$$

3. Find all primitive Pythagorean triangles whose perimeter is less than 100. [HINT: There are 7 such triangles. Show that for such a triangle, $m(m + n) < 50$ and hence $m^2 + m < 50$ and thus $m \leqslant 6$.]

4. Prove that the perimeter of a Pythagorean triangle divides twice its area.

5. Find a Pythagorean triangle whose hypotenuse is 73.

Building Pythagorean triangles on a given side

Every number $s \geqslant 3$ can be a side of a Pythagorean triangle. For if s is odd, say $s = 2n + 1$, then s is a side of the Pythagorean triangle $T(2n + 1, 2n(n + 1), 2n^2 + 2n + 1)$ corresponding to $k = 1$ and $m = n + 1$. If s is even, say $s = 2m$, then s is a side of the Pythagorean triangle $T(m^2 - 1, 2m, m^2 + 1)$ corresponding to $k = 1$, $n = 1$.

Many numbers cannot be the hypotenuse of a Pythagorean triangle. There is, for example, no Pythagorean triangle with an hypotenuse of length 57.

Instead of looking at the general question of Pythagorean hypotenuses, let us ask the narrower question: 'which *prime* numbers can be the hypotenuse of a Pythagorean triangle?'

Suppose that the prime p is the hypotenuse of a Pythagorean triangle. Then p has the form

$$p = k(m^2 + n^2) \text{ where } m > n > 0.$$

Since $m^2 + n^2 > 1$ and p is prime, k must be 1. So,

$$p = m^2 + n^2.$$

Since $m > n$, $p \neq 2$. Now n lies strictly between 0 and p and since p is prime, n and p must be coprime. Thus by the theory in chapter 1, the equation

$$nx \equiv 1 \; (\mathrm{mod}\; p)$$

has a solution. Squaring, we obtain

$$n^2 x^2 \equiv 1 \; (\mathrm{mod}\; p).$$

But
$$(m^2 + n^2)\, x^2 \equiv px^2 \equiv 0 \; (\mathrm{mod}\; p)$$

and so, subtracting,

$$m^2 x^2 \equiv -1 \; (\mathrm{mod}\; p).$$

Thus -1 is a quadratic residue modulo p, i.e. $(-1|p) = 0$. Using the theory from chapter 2, we conclude that

$$\frac{p-1}{2} \equiv 0 \; (\mathrm{mod}\; 2)$$

and so
$$p \equiv 1 \; (\mathrm{mod}\; 4).$$

We have shown that if a prime is the hypotenuse of a Pythagorean triangle, it must be of the form $4k + 1$ (these are the primes in blue in the table of primes on page 34.) Conversely it can be shown that every prime of the form $4k + 1$ can be written in the form $m^2 + n^2$ where $m > n$ and hence it is the hypotenuse of $T(m^2 - n^2, 2mn, m^2 + n^2)$.

Using similar methods to these it is possible to prove the more general fact that:

> **a positive number is the hypotenuse of a Pythagorean triangle if and only if it is divisible by at least one prime of the form $4k + 1$.**

Example. Which of the following is the hypotenuse of a Pythagorean triangle? (a) 57, (b) 58.
For any which are, find a suitable triangle.

Solution. (a) $57 = 3 \times 19$. Neither 3 nor 19 is congruent to 1 modulo 4 and so 57 cannot be the hypotenuse of a Pythagorean triangle.

(b) $58 = 2 \times 29$. Since $29 \equiv 1 \; (\mathrm{mod}\; 4)$, 58 is the hypotenuse of a Pythagorean triangle. Since

$$29 = 5^2 + 2^2,$$

29 is the hypotenuse of $T(5^2 - 2^2, 2 \times 10, 5^2 + 2^2)$, i.e. $T(21, 20, 29)$. Hence 58 is the hypotenuse of $T(42, 40, 58)$.

Throughout this chapter we have been concerned with solutions to the equation

$$x^2 + y^2 = z^2.$$

We might ask: 'what are the positive solutions of the equation

$$x^3 + y^3 = z^3?'$$

The answer is that there are none. Likewise, the equation

$$x^4 + y^4 = z^4$$

has no positive solutions.

Fermat, a great seventeenth-century mathematician, made a practice of scribbling comments in the margins of his books. One such comment was the following:

'. . . it is impossible to write a cube as the sum of two cubes, and a fourth power as the sum of two fourth powers and in general any power beyond the second as the sum of two similar powers. For this I have discovered a truly wonderful proof, but the margin is too small to contain it.'

What Fermat claimed to have proved is that the equation

$$x^n + y^n = z^n$$

has no positive solutions if $n > 2$. Did Fermat possess a proof of this? We do not know. He made *conjectures* which turned out to be false (e.g. that $2^{2^n} + 1$ is always prime) but every other fact he claimed to have *proved* has turned out to be true.

We do not even know if the statement is in fact true. For the last three hundred years some of the best mathematical brains have tackled this problem without finally settling it. No positive solution to such an equation has ever been found for any $n > 2$. Nobody has been able to prove that there are none. The statement has indeed been proved for values of n up to several millions, but never for *all* $n > 2$.

In 1908 a German mathematician left 100,000 marks, then quite a fortune, to be awarded to the first person who proves the so-called 'Last Theorem of Fermat' and although thousands of 'proofs' have been submitted, mostly by amateurs, none were correct. The 'fortune' was reduced to almost nothing by post-war inflation, but the problem still attracts the attention of not only professional mathematicians and amateurs, but also of cranks who delude themselves that they have a proof that nobody else will accept.

1. Find Pythagorean triangles which include the following numbers as the length of a side:
   (a) 17, (b) 52.

2. Which of the following primes are the length of the hypotenuse of a Pythagorean triangle?
   (a) 37, (b) 41, (c) 43.

3. Which of the following numbers are the length of the hypotenuse of a Pythagorean triangle?
   (a) 90, (b) 91, (c) 92, (d) 93, (e) 94.

4. Find a Pythagorean triangle which has 106 as the length of its hypotenuse.

5. Prove that there is no Pythagorean triangle whose sides are all perfect squares. [HINT: Use Fermat's Last Theorem (for one of the values of n for which it has been proved)]

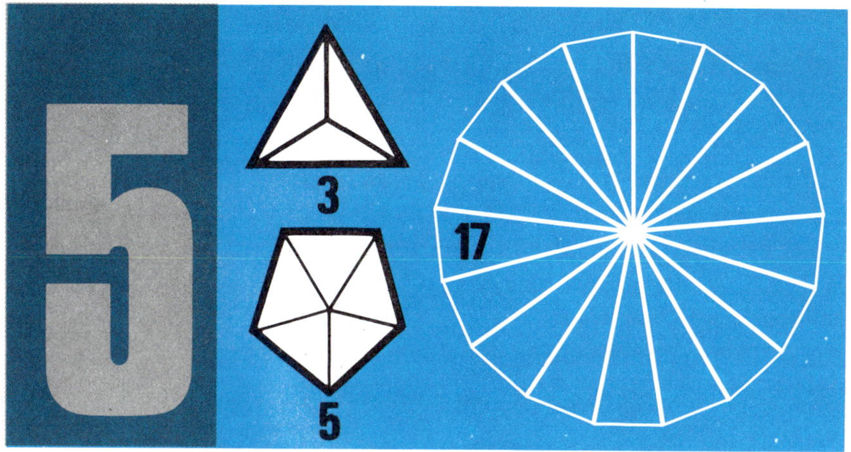

Fermat Primes and Euler's Phi-function

'Good evening. My number is $\phi(v)$, pronounced 'phi v', or 'five'. I would like to speak to you about a rather serious matter that we must all face sooner or later. Have you ever given any thought to the inscription you would like on your tombstone? May I suggest that an odd-sided regular polygon would be a distinctive yet dignified memorial to you.

'The Constructible Polygon Company has constructed odd-sided polygons now for many generations using only the highest quality rulers and compasses. The work is carried out by five highly-skilled Fermat primes: 3 our managing director, 17, 257, 65537 and myself.

'Both 3 and I have been in the business for many centuries, but our range of sizes was for a long time very small — just the 3-sided, 5-sided and 15-sided polygons. Then at the end of the eighteenth century, the mathematician Gauss arranged for some other Fermat primes to join us. This enabled us to extend our range considerably and now you can choose from 31 sizes, ranging from the 3-sided to the 4294967295-sided one.

'Most customers prefer the 5-sided or the 17-sided polygons. The 17-sided one, for example, was inscribed on the monument to Gauss in his home town.'

Euler's phi-function

The number of numbers between 1 and m inclusive which are coprime with m, is denoted by $\phi(m)$.

Example. Find $\phi(12)$.

Solution. Of the numbers 1, 2, 3, 4, 5, 6, 7, 8, 9, 10, 11 and 12 the only ones which are coprime with 12 are 1, 5, 7 and 11. Hence $\phi(12) = 4$.

If p is any prime number, the only number from 1 to p which is *not* coprime with p is p itself. So $\phi(p) = p - 1$.

Let us now consider $\phi(p^r)$ for some prime p. Which of the numbers $1, 2, 3 \ldots p^r$ are *not* coprime with p^r? The answer is clearly 'all of the multiples of p', viz. $p, 2p \ldots p^{r-1}p$. There are p^{r-1} of them. This leaves $p^r - p^{r-1}$ numbers which *are* coprime with p^r. Hence

$$\phi(p^r) = p^r - p^{r-1} = p^{r-1}(p - 1).$$

Now let us investigate $\phi(p^r q^s)$ where p and q are primes. The numbers between 1 and $p^r q^s$ inclusive which are *not* coprime with $p^r q^s$ are the multiples of p and the multiples of q. Now there are $p^{r-1}q^s$ multiples of p and $p^r q^{s-1}$ multiples of q. We cannot simply add these together otherwise some numbers would be counted twice. Which numbers are *both* a multiple of p and a multiple of q? The multiples of pq. How many are there? There are $p^{r-1}q^{s-1}$ multiples of pq, viz. $pq\ 2pq \ldots p^{r-1}q^{s-1}(pq)$. Hence the number of numbers between 1 and $p^r q^s$ inclusive which are not coprime with $p^r q^s$ is

$$p^{r-1}q^s + p^r q^{s-1} - p^{r-1}q^{s-1}$$

and so the number which *are* coprime with $p^r q^s$ is

$$\begin{aligned}
\phi(p^r q^s) &= p^r q^s - (p^{r-1}q^s + p^r q^{s-1} - p^{r-1}q^{s-1}) \\
&= p^r q^s - p^{r-1}q^s - p^r q^{s-1} + p^{r-1}q^{s-1} \\
&= (p^r - p^{r-1})(q^s - q^{s-1}) \\
&= p^{r-1}(p - 1)\,q^{s-1}(q - 1) \\
&= \phi(p^r)\,\phi(q^s).
\end{aligned}$$

By a similar argument we can show that this extends to the following general formula:

$$\phi(p_1{}^{r_1}\,p_2{}^{r_2} \ldots p_k{}^{r_k}) = \phi(p_1{}^{r_1})\,\phi(p_2{}^{r_2}) \ldots \phi(p_k{}^{r_k})$$

where $p_1, p_2, \ldots, p_k$ are different primes

Example
$$\begin{aligned}
\phi(600) &= \phi(2^3 \times 3 \times 5^2) \\
&= \phi(2^3)\,\phi(3)\,\phi(5^2) \\
&= 2^2\,(2 - 1)\,3^0\,(3 - 1)\,5^1\,(5 - 1) \\
&= 2^2 \times 2 \times 20 \\
&= 160
\end{aligned}$$

Constructible polygons

The reader is probably familiar with certain geometric constructions
which can be carried out using ruler and compass, such as the bisection
of a line, bisection of an angle, construction of a perpendicular to a
given line etc. Many geometric figures can be constructed using only a
ruler and compass, such as a circle, a 45°, 45°, 90° triangle, a 30°,
60°, 90° triangle, rectangles etc.

A *regular n-sided polygon* is a figure bounded by n straight sides
whose vertices are equally spaced around a circle. A regular 3-sided
polygon is an equilateral triangle and a regular 4-sided polygon is a
square. The following diagrams indicate how regular 3-sided, 4-sided,
5-sided and 6-sided polygons can be constructed by ruler and compass.

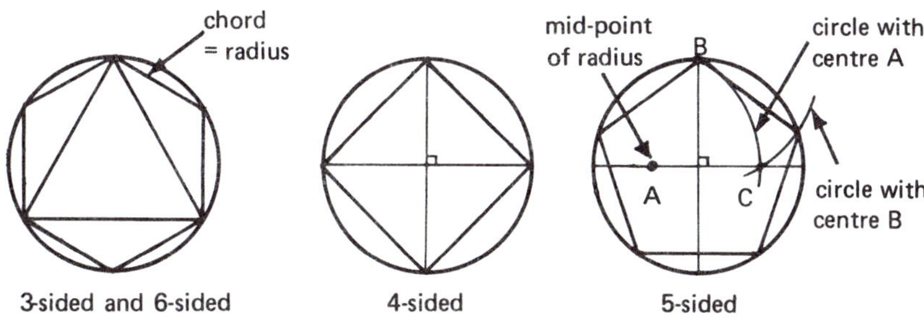

The Greeks knew how to construct regular polygons with 2^n, $3 \times
2^n$, 5×2^n and 15×2^n sides. It was not until the end of the eighteenth
century that the teenage Gauss constructed a regular 17-sided polygon.

He proved, moreover, that a regular n-sided polygon can be constructed
by ruler and compass if, and only if, $\phi(n)$ is a power of 2. For example
$\phi(3) = 2$, $\phi(4) = 2$, $\phi(5) = 4$, $\phi(6) = 2$, $\phi(15) = \phi(3)\,\phi(5) = 8$,
$\phi(17) = 16$.

When is $\phi(m)$ a power of 2?

For what numbers, m, is $\phi(m)$ a power of 2? Suppose that

$$m = p_1^{r_1}\, p_2^{r_2} \ldots \ldots p_k^{r_k}$$

is the factorisation of m into different primes $p_1, p_2, \ldots \ldots, p_k$. Then

$$\phi(m) = \phi(p_1^{r_1})\, \phi(p_2^{r_2}) \ldots \phi(p_k^{r_k}).$$

If $\phi(m)$ is a power of 2, each $\phi(p_i^{r_i})$ must also be a power of 2.

When is $\phi(p^r)$ a power of 2? We know that

$$\phi(p^r) = p^{r-1}\,(p - 1).$$

If $p = 2$, this becomes 2^{r-1} and is clearly a power of 2. If p is odd, then for $p^{r-1}(p-1)$ to be a power of 2, r must be equal to 1 (otherwise p would divide $\phi(p^r)$). Moreover, $p-1$ must be a power of 2. That is, we must have

$$p = 1 + 2^n$$

for some n. As we have seen, such primes are called Fermat Primes and n must itself be a power of 2.

So an m-sided regular polygon is constructible by ruler and compass if and only if m has the form

$$2^r p_1 p_2 \ldots p_k$$

where $p_1, p_2, \ldots, p_k$ are different Fermat primes.

Since the only known Fermat primes are 3, 5, 17, 257 and 65537, the only known odd-sided regular polygons which are constructible by ruler and compass are those with 3, 5, 17, 257, 65537, 3 × 5, 3 × 17, 5 × 17, ..., 3 × 5 × 17, ... sides, and there are just 31 of these. The largest known odd number for which a regular polygon can be constructed is 4294967295 = 3 × 5 × 17 × 257 × 65537.

Gauss's proof shows that regular polygons for values of m which do not have the above form cannot be constructed by ruler and compass. That is, there is no way a regular 7-sided polygon, for example, can be constructed by ruler and compass (though it is possible to construct a good approximation to one).

There is a story that Gauss suggested that the regular 17-sided polygon should be inscribed on his gravestone. This was not done, but instead it appears on the monument in his home town, Brunswick.

EXERCISE SET 12

1. Find $\phi(n)$ for each of the following values of n:
   (a) 91, (b) 92, (c) 93, (d) 94, (e) 95.

2. Find as many values of n as possible for which $\phi(n) = 24$.

3. Which odd-sided regular polygons with fewer than 1000 sides can be constructed by ruler and compass?

4. (Harder) For what values of m is $\phi(m)$ a power of 3?

Perfect Numbers

'I know this will sound like I'm boasting, but I am a perfect number.
You see my divisors, excluding myself, are 1, 2 and 3 and $1 + 2 + 3 = 6$.
That's perfection. My sister, 28, is also perfect. Her divisors are 1, 2, 4,
7 and 14 (not counting herself) and $1 + 2 + 4 + 7 + 14 = 28$. You see.
That's also perfection. The fact that the world was created in 6 days
and that the moon takes 28 days to revolve around the earth . . . these
aren't accidents you know. The creation was perfect and so it had to
be done with perfect numbers.

'I take no credit for my perfection — I was born with it. You see,
my father was the Mersenne prime 3, and we perfect numbers are all
descended from Mersenne primes. That is why as yet there are so few
of us. Only about two dozen. If ever another prime discovers that he's
Mersenne another perfect number will be brought into the world. She'll
be an even number like the rest of us. That is not to say that an odd
perfect number might not turn up one day, but if he does exist he
must be a long, long way from here.'

Perfection

A *perfect number* is defined to be one which is equal to the sum of its
proper divisors (that is, excluding itself). The first three perfect numbers
are

$$6 = 1 + 2 + 3,$$
$$28 = 1 + 2 + 4 + 7 + 14,$$

and $$496 = 1 + 2 + 4 + 8 + 16 + 31 + 62 + 124 + 248.$$

Perfect numbers have fascinated men throughout the ages and have had supernatural significances attributed to them. The number 6 has been described as 'marriage and health and beauty on account of the integrity of its parts and the agreement existing in it' and as 'a symbol of the union of the two sexes, that is, from three which is male and from two, which is feminine since it is even.'

Euclid, or one of his followers, discovered the following formula for perfect numbers:

$$2^{n-1} (2^n - 1)$$

where $2^n - 1$ is prime. We saw in chapter 3 that such a prime is called a Mersenne prime and that n is itself prime. So, if $p = 2^n - 1$ is any Mersenne prime then

$$\frac{p (p + 1)}{2}$$

is a perfect number.

Table 3 : The first six perfect numbers

n	Mersenne prime = $2^n - 1$	Perfect Number = $2^{n-1} (2^n - 1)$
2	3	6
3	7	28
5	31	496
7	127	8128
13	8191	33550336
17	131071	8589869056

For any number, n, we define $s(n)$ to be the sum of the divisors of n excluding n itself, and $S(n)$ to be the sum of all of the divisors of n (including n itself). They are clearly related by the equation

$$S(n) = s(n) + n.$$

Example

$S(2) = 1 + 2 = 3$ $s(2) = 1$

$S(3) = 1 + 3 = 4$ $s(3) = 1$

$S(4) = 1 + 2 + 4 = 7$ $s(4) = 1 + 2 = 3$

$S(6) = 1 + 2 + 3 + 6 = 12$ $s(6) = 1 + 2 + 3 = 6$

If n is perfect then $s(n) = n$ and $S(n) = 2n$. If p is prime,

$$S(p^r) = 1 + p + p^2 + \ldots + p^r$$
$$= \frac{p^{r+1} - 1}{p - 1}$$

using the formula for the sum of a geometric progression.

Suppose that p is a prime dividing n and that

$$n = p^r k$$

where p does not divide k (i.e. we take as many factors of p as possible out of n). Let the divisors of k be 1, k and $d_1 \ldots d_t$. Then

$$
\begin{aligned}
S(n) = &\ 1 + p + p^2 + \ldots + p^r \\
&+ d_1 + d_1\, p + d_1 p^2 + \ldots + d_1 p^r \\
&+ \ldots \ldots \ldots \ldots \ldots \ldots \ldots \\
&+ d_t + d_t p + d_t p^2 + \ldots + d_t p^r \\
&+ k + kp + kp^2 + \ldots + kp^r \\
= &\ (1 + p + p^2 + \ldots + p^r)\,(1 + d_1 + d_2 + \ldots + d_t + k) \\
= &\ S(p^r)\, S(k).
\end{aligned}
$$

By a repeated application of this we obtain the theorem:

if $n = p_1^{r_1} p_2^{r_2} \ldots p_m^{r_m}$ is the factorisation
of n into different primes then
$$S(n) = S(p_1^{r_1})\, S(p_2^{r_2}) \ldots S(p_m^{r_m}).$$

We shall now check that Euclid's formula always gives perfect numbers. If $p = 2^r - 1$ is prime and $n = 2^{r-1} p$, then

$$
\begin{aligned}
S(n) = S(2^{r-1} p) &= S(2^{r-1})\, S(p) \\
&= (2^r - 1) \cdot (p + 1) \\
&= p\, 2^r \\
&= 2n,
\end{aligned}
$$

and so n is perfect.

No other even perfect numbers

We now show that every even perfect number is given by Euclid's formula. Suppose that n is an even perfect number and let

$$n = 2^r k$$

where $r \geqslant 1$ and k is odd. Then

$$
\begin{aligned}
S(n) = S(2^r)\, S(k) \\
= (2^{r+1} - 1)\, S(k).
\end{aligned}
$$

Since n is perfect,

$$S(n) = 2n = 2^{r+1} k.$$

Hence $\qquad (2^{r+1} - 1)\, S(k) = 2^{r+1} k$

and so $\qquad (2^{r+1} - 1)\, (s(k) + k) = 2^{r+1} k$

whence $\qquad (2^{r+1} - 1)\, s(k) = k.$

Since $2^{r+1} - 1$ is bigger than 1, $s(k)$ is a proper divisor of k. But $s(k)$ is the *sum* of the proper divisors of k. Hence k has only one proper divisor, viz. 1. That is, k is prime and $s(k) = 1$. Hence $k = 2^{r+1} - 1$ and so

$$n = 2^r\,(2^{r+1} - 1)$$

where $2^{r+1} - 1$ is a Mersenne prime.

The existence of odd perfect numbers is still an unsettled question. We do not know if there are any or not, though we do know that there are none with fewer than 15 digits.

EXERCISE SET 13

1. Find the following:
   (a) s(220);   (b) s(284);   (c) s(1184);   (d) s(1210).

2. Find the following:
   (a) S(120);   (b) S(30240).

3. (Harder) Prove that the only even numbers, n, such that $S(n)$ is prime, are those of the form 2^r where $2^r - 1$ is a Mersenne prime.

Repeating Decimals

'I am number 7 and this is my reciprocal, 1/7. Don't let his barking frighten you. He's quite a friendly reciprocal really. You see, he's wagging his tail. Now it is this long tail that I want to talk to you about. It's a six-digit tail. That is,

$$1/7 = 0.142857\ 142857\ldots\ldots$$

and the repeating part, or tail, is 6 digits long.

'Now all numbers have reciprocals. All except for 0, that is. She's too stupid to look after one. And every other number here on the platform has a reciprocal with a short tail, just 1 digit long. Hardly long enough for the poor reciprocal to wag.

'I have made a study of reciprocals' tails and have discovered a way of calculating the length of the tail of any reciprocal just from knowing the number of its master. I might tell you about it some day.'

Decimal expansions

The decimal expansion of any *rational number* (that is, a number of the form m/n where m and n are whole numbers) reaches a point where a group of digits is repeated over and over again. For example,

$$11/14 = 0.7\ 857142\ 857142\ 857142\ldots\ldots$$

which we usually write as $0.7\dot{8}5714\dot{2}$, where the dots above the 8 and 2 indicate that this section of the expansion is repeated indefinitely. In contrast to this, *irrational numbers,* such as $\sqrt{2}$ and π, do not have repeating decimal expansions.

The number of digits in this repeating part is called the *period* of the expansion. To gain some insight into the way this is related to the rational number, we shall calculate the decimal expansion of 2/7.

$$
\begin{array}{r}
0.2857142 \\
7\overline{\smash{\big)}\,2.0000000} \\
1.4 \\
\hline
6\,0 \\
5\,6 \\
\hline
4\,0 \\
3\,5 \\
\hline
5\,0 \\
4\,9 \\
\hline
1\,0 \\
7 \\
\hline
3\,0 \\
2\,8 \\
\hline
2\,0 \\
1\,4 \\
\hline
6 \quad \text{etc.}
\end{array}
$$

Thus $2/7 = 0.\dot{2}8571\dot{4}$.

At each stage the remainder completely determines the rest of the computation. The digits in the decimal expansion begin to repeat when the remainders begin to repeat.

Now the remainder at the first step is the remainder on dividing 20 by 7, the remainder at the second step is the remainder on dividing 200 by 7 and so on. The remainder at the seventh step is the remainder on dividing 2×10^7 by 7. Both 20 and 2×10^7 give a remainder of 6 on dividing by 7. Hence

$$2 \times 10^7 \equiv 20 \ (\text{mod } 7).$$

Since 20 is coprime with 7, we may divide through by 20 to obtain

$$10^6 \equiv 1 \ (\text{mod } 7).$$

In fact 10^6 is the first power of 10 which is congruent to 1 modulo 7 (after $10^0 = 1$). If an earlier power had been congruent to 1 modulo 7, the decimal expansion would have begun to repeat earlier.

In general, if the decimal expansion for the rational number m/n has period r and if there are s decimal places before the repeating part, then r and s are the smallest numbers such that

$$10^{r+s} m \equiv 10^s m \pmod{n}.$$

If m/n has been 'reduced to lowest terms' (i.e. no 'cancelling' is possible), m and n will be coprime. We may therefore divide through by m to obtain

$$10^{r+s} \equiv 10^s \pmod{n}.$$

Thus if m/n has been reduced to lowest terms, the numbers r and s depend only on the denominator, n.

If n is coprime with 10, we may divide through by 10^s to obtain

$$10^r \equiv 1 \pmod{n}.$$

Example. Find the period of 3/26. At what point does the decimal expansion begin to repeat?

Solution. Modulo 26,

$$
\begin{aligned}
10^0 &\equiv 1 \\
10^1 &\equiv 10 \\
10^2 &\equiv 100 \equiv 22 \\
10^3 &\equiv 220 \equiv 12 \\
10^4 &\equiv 120 \equiv 16 \\
10^5 &\equiv 160 \equiv 4 \\
10^6 &\equiv 40 \equiv 14 \\
10^7 &\equiv 140 \equiv 10
\end{aligned}
$$

Hence the smallest numbers r and s such that

$$10^{r+s} \equiv 10^s \pmod{26}$$

are $r = 6$, $s = 1$. The decimal expansion of 3/26 thus has a period of 6 and there is one decimal place before the repeating part begins. (N.B. $3/26 = 0.1\dot{1}5384\dot{6}$)

Example. Find the period of 35/111. At what point does the decimal expansion begin to repeat?

Solution. Modulo 111, $10^0 \equiv 1$, $10^1 \equiv 10$, $10^2 \equiv 100$, $10^3 \equiv 1000 \equiv 1$. Hence the smallest numbers r and s such that $10^{r+s} \equiv 10^s \pmod{111}$ are $r = 3$, $s = 0$ and so there are no decimal places before the repeating part and the period is 3. (N.B. $35/111 = 0.\dot{3}1\dot{5}$.)

Exponents

The smallest positive number, n, such that

$$a^n \equiv 1 \pmod{m}$$

is called the *exponent* of a modulo m. It exists only if $(a, m) = 1$ for if $(a, m) > 1$, a^n will never be congruent to 1 modulo m.

The period of m/n (if n is coprime to both m and 10) is thus the exponent of 10 modulo n.

Suppose that n is the exponent of a modulo m and that N is any number such that

$$a^N \equiv 1 \pmod{m}.$$

Divide N by n and let the quotient be q and the remainder be r. Then

$$N = qn + r, \text{ and } r < n.$$

Hence

$$a^N = a^{qn+r} = (a^n)^q a^r.$$

Since $a^n \equiv 1 \pmod{m}$ and $a^N \equiv 1 \pmod{m}$ we have

$$1 \equiv a^r \pmod{m}.$$

But $r < n$, and n is the *smallest* positive number such that $a^n \equiv 1 \pmod{m}$. Hence r cannot be positive. It must therefore be zero and so $N = qn$. Therefore

> **the exponent of a modulo m divides any number N such that $a^N \equiv 1 \pmod{m}$.**

A famous theorem of Euler states that

> **if a and m are coprime, $a^{\phi(m)} \equiv 1 \pmod{m}$**

where ϕ is the Euler ϕ-function which was introduced in Chapter 5. We may therefore conclude that if a and m are coprime,

> **the exponent of a modulo m divides $\phi(m)$.**

Example. Find the exponent of 2 modulo 21 and verify that it divides $\phi(21)$.

Solution. $2^6 = 64 \equiv 1 \pmod{21}$. Hence the exponent of 2 modulo 21 divides 6. However neither 2^1, nor 2^2 nor 2^3 is congruent to 1 modulo 21. Hence the exponent of 2 modulo 21 is 6. Now $\phi(21) = \phi(3)\phi(7) = 2 \times 6 = 12$.

A consequence of Euler's theorem is that

**if n is coprime with both 10 and m, the
period of the decimal expansion of m/n divides $\phi(n)$.**

Examples.

The period of 2/7 is 6 and $\phi(7) = 6$.
The period of 3/26 is 6 and $\phi(26) = \phi(2)\phi(13) = 12$.
The period of 35/111 is 3 and $\phi(111) = \phi(3)\phi(37) = 72$.

Example. Find the period of 1/107.

Solution. The period of 1/107 is the exponent of 10 modulo 107. It
divides $\phi(107) = 106$ (since 107 is prime) $= 2 \times 53$. It must therefore
be either 1, 2, 53 or 106. But neither 10^1 nor 10^2 is congruent to 1
modulo 107. Hence the exponent of 10 modulo 107 is either 53 or 107.
We now compute 10^{53}, modulo 107.

$$\begin{aligned}
10^2 \quad &= 100 \equiv -7 \\
10^4 \quad &\equiv (-7)^2 \equiv 49 \\
10^5 \quad &\equiv 490 \equiv 62 \equiv -45 \\
10^{10} &\equiv (-45)^2 \equiv 2025 \equiv -8 \\
10^{20} &\equiv (-8)^2 \equiv 64 \equiv -43 \\
10^{40} &\equiv (-43)^2 \equiv 1849 \equiv 30 \\
10^{50} &\equiv (-8) \times 30 \equiv -240 \equiv -26 \\
10^{52} &\equiv (-7) \times (-26) \equiv 182 \equiv 75 \\
10^{53} &\equiv 750 \equiv 1.
\end{aligned}$$

Hence the exponent of 10 modulo 107 is 53 and so 1/107 has a decimal
expansion with period 53. If 10^{53} had not been congruent to 1 modulo
107, the exponent of 10 modulo 107 would have been 106. No further
calculation would have been necessary.

A simpler solution to this problem makes use of the theory of
quadratic residues (see chapter 2). Having decided that the exponent of
10 modulo 107 is either 53 or 106 we can argue thus: suppose that it
is 106. Then $10^{106} \equiv 1 \pmod{107}$, but $10^{53} \not\equiv 1 \pmod{107}$. Hence 107
divides $10^{106} - 1$ which is equal to $(10^{53} - 1)(10^{53} + 1)$, but does not
divide $10^{53} - 1$. Therefore 107 divides $10^{53} + 1$ and so $10^{53} \equiv -1$
$\pmod{107}$, whence $10^{54} \equiv -10 \pmod{107}$. Therefore -10 is a quadratic
residue modulo 107, i.e. $(-10|107) = 0$. However, if $(-10|107)$ is
computed using the methods of chapter 2, its value is seen to be 1.
Hence the exponent of 10 modulo 107 cannot be 106 and therefore it
must be 53.

By a similar argument, one can show that if p is any prime of the form $2q + 1$, where q is also prime, and if p does not divide a, and if $(-a|p) = 1$, the exponent of a modulo p is equal to 2 or q. The proof is left to the reader.

EXERCISE SET 14

1. Find the exponents of 2 modulo n for the following values of n:
   (a) 7; (b) 9; (c) 11.

2. Find the period of the decimal expansions of the following:
   (a) 1/37; (b) 1/83; (c) 1/139.

3. (Harder) Prove that if p and q are primes of the form $q = 20k + 1$, $p = 40k + 3$ then the decimal expansion of $1/p$ has period q.
   [HINT: Use quadratic residues]

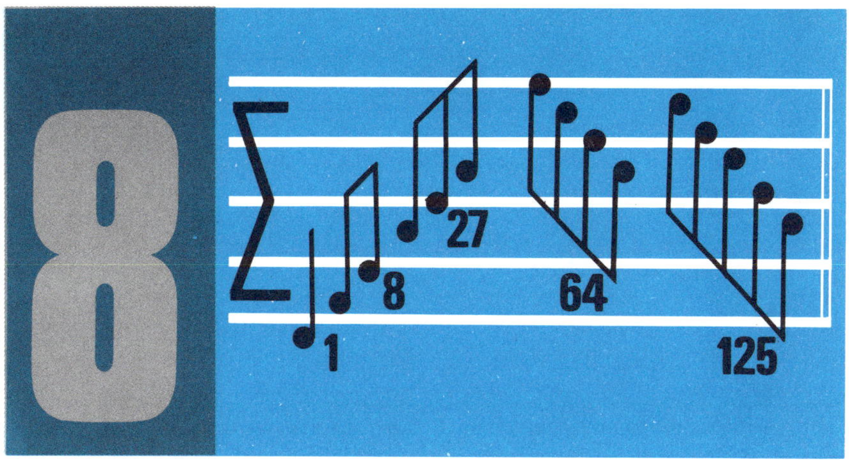

Sums of Cubes and Other Sums

'My real name is 8, though I'm perhaps better known by my professional name, 2^3, and I would like to sum for you a sum entitled 'The Eight Days of Cubefeast'. (No, that note's too high. Could I have it an octave lower? Thank-you.)

'On the first day of Cubefeast my true-love gave to me
   A partridge in a pear tree.
On the second day of Cubefeast my true-love gave to me
   8 turtle doves and a partridge in a pear tree.
. . . .
On the eighth day of Cubefeast my true-love gave to me
   512 ladies singing,
   343 children playing,
   216 maids a-dancing,
   125 golden rings;
    64 calling birds,
    27 French hens,
     8 turtle doves
       and a partridge in a pear tree.
Throughout the time of Cubefeast my true-love gave to me,
   36^2 gifts on day 8,
   28^2 gifts on day 7,
   21^2 gifts on day 6,
   15^2 on the 5th;
   10^2 – day 4,

$$6^2 - \text{three,}$$
$$3^2 \text{ on two,}$$

plus the partridge on the first day.'

Sigma notation

A convenient notation for a sum of terms such as

$$a_1 + a_2 + \ldots + a_n$$

is

$$\sum_{r=1}^{n} a_r.$$

This means that the typical term in the sum is a_r and that in the first term, $r = 1$ and in the last term, $r = n$. If, as will often be the case throughout this chapter, the value of n is understood, we may simply write

$$\sum a_r.$$

Examples
$$1 \times 2 \times 3 + 2 \times 3 \times 4 + \ldots + n(n+1)(n+2)$$
$$= \sum r(r+1)(r+2).$$
$$1^3 + 2^3 + \ldots + n^3 = \sum r^3.$$
$$1 + 1 + \quad + 1 = \sum 1.$$

Some simple properties of sums in the sigma notation are

$$\sum (a_r + b_r) = \sum a_r + \sum b_r$$

and
$$\sum ka_r = k \sum a_r$$

where k is a constant.

Example.
$$\sum r(r+1) = \sum (r^2 + r) = \sum r^2 + \sum r.$$

Sums of products of successive numbers

We shall now obtain a formula for

$$\sum_{r=1}^{n} r(r+1)(r+2).$$

Put
$$a_r = r(r+1)(r+2)(r+3).$$

Then
$$a_r - a_{r-1} = r(r+1)(r+2)(r+3) - (r-1)r(r+1)(r+2)$$
$$= r(r+1)(r+2)[(r+3) - (r-1)]$$
$$= 4r(r+1)(r+2).$$

Thus

$$4 \sum r(r + 1)(r + 2) = \sum 4r(r + 1)(r + 2)$$
$$= \sum (a_r - a_{r-1})$$
$$= (a_1 - a_0) + (a_2 - a_1) + (a_3 - a_2) + \ldots$$
$$\ldots + (a_n - a_{n-1})$$
$$= a_n - a_0$$
$$= a_n, \text{ since } a_0 = 0.$$

Hence

$$\sum_{r=1}^{n} r(r + 1)(r + 2) = \frac{1}{4} a_n = \frac{1}{4} n(n + 1)(n + 2)(n + 3).$$

Example

$$1 \times 2 \times 3 + 2 \times 3 \times 4 + \ldots + 8 \times 9 \times 10$$
$$= \frac{1}{4} \times 8 \times 9 \times 10 \times 11 = 1980.$$

In a similar way, one can show that

$$\sum r(r + 1) = \frac{1}{3} n(n + 1)(n + 2),$$

$$\sum r = \frac{1}{2} n(n + 1),$$

and, in general, that

$$\sum_{r=1}^{n} r(r + 1)(r + 2) \ldots (r + k - 1)$$

$$= \frac{1}{k + 1} n(n + 1)(n + 2)(n + 3) \ldots (n + k).$$

Sums of squares

To find $\sum r^2$ we express r^2 in terms of $r(r + 1)$.

$$\sum_{r=1}^{n} r^2 = \sum [r(r + 1) - r]$$
$$= \sum r(r + 1) - \sum r$$
$$= \frac{1}{3} n(n + 1)(n + 2) - \frac{1}{2} n(n + 1)$$
$$= \frac{1}{6} n(n + 1)(2n + 4 - 3)$$
$$= \frac{1}{6} n(n + 1)(2n + 1).$$

Example. A pile of cannon balls is stacked in such a way that the rth layer from the top forms a square of side r. If there are 24 layers, how many cannon balls are there altogether?

Solution. Total number of cannon balls

$$= \sum_{r=1}^{24} r^2$$

$$= \frac{1}{6} \times 24 \times 25 \times 49 = 4900.$$

Sums of cubes

$$\sum_{r=1}^{n} r^3 = \sum [r(r + 1)(r + 2) - 3r^2 - 2r]$$

$$= \sum r(r + 1)(r + 2) - 3 \sum r^2 - 2 \sum r$$

$$= \frac{1}{4} n(n + 1)(n + 2)(n + 3) - \frac{3}{6} n(n + 1)(2n + 1) - \frac{2}{2} n(n + 1)$$

$$= \frac{1}{4} n(n + 1)[(n + 2)(n + 3) - 2(2n + 1) - 4]$$

$$= \frac{1}{4} n(n + 1)(n^2 + n)$$

$$= \frac{1}{4} n^2 (n + 1)^2 = (\Sigma r)^2.$$

It is interesting to note that although Σr^3 is always a perfect square, it can be shown that Σr^2 is a perfect square only if $n = 1$ or 24.

Example. How many gifts did 8's true-love give her on the eighth day according to the song at the beginning of the chapter?

Solution. Number of gifts on the eighth day

$$= \sum_{r=1}^{8} r^3$$

$$= (\frac{1}{2} \times 8 \times 9)^2 = 36^2 = 1296$$

Example. In the song 'The Twelve Days of Christmas', the singer gets $1 + 2 + \ldots + r$ gifts on the rth day. How many gifts does she get altogether?

Solution. Total number of gifts $= \displaystyle\sum_{r=1}^{12} \left[\frac{1}{2} r(r+1)\right]$

$$= \frac{1}{2}\left(\sum_{r=1}^{12} r(r+1)\right)$$

$$= \frac{1}{6} \times 12 \times 13 \times 14 = 364.$$

That is, she gets a gift for every day of the year except Christmas!

1. Find $1 \times 3 + 2 \times 4 + \ldots + 100 \times 102$.

2. Find $1 \times 2 + 3 \times 4 + \ldots + 25 \times 26$.
   [HINT: General Term is $(2r - 1)(2r)$]

3. Find $1 + 3 + 5 + \ldots + 2n - 1$.

4. Find a formula in terms of n for $\displaystyle\sum_{r=1}^{n} r^4$.

   [HINT: $r^4 = r(r+1)(r+2)(r+3) - 6r^2 - 11r - 6$]

5. How many gifts did 8 receive altogether during the eight days of Cubefeast?

Waring's problem

Waring's problem is concerned with expressing a number n in the form

$$n = x_1{}^k + \ldots + x_r{}^k$$

where $k \geqslant 2$ and each $x_i \geqslant 0$. Clearly every positive integer n can be
so expressed:

$$n = 1^k + \ldots + 1^k.$$

However, there are usually shorter expressions than this. For example,
91 can be expressed as a sum of cubes in the following ways:

$$91 = 1 + 1 + 1 + 1 + 1 + 8 + 8 + 8 + 8 + 27 + 27,$$
$$91 = 1 + 1 + 1 + 8 + 8 + 8 + 64,$$
$$91 = 1 + 1 + 8 + 81,$$
$$91 = 27 + 64.$$

The least number of positive kth powers required for a number n can
be denoted by $W_k(n)$. Thus $W_3(91) = 2$ since 91 can be expressed as a
sum of two positive cubes but no fewer. If n is itself a cube, $W_3(n) = 1$.

As a final example, $W_3(23) = 9$ since the shortest expression of 23 as a sum of positive cubes is

$$23 = 1 + 1 + 1 + 1 + 1 + 1 + 1 + 8 + 8.$$

Now one might expect that although $W_3(n)$ can be small for large values of n, in general it would increase with n. The surprising fact is, however, that for a fixed value of k, there is an upper limit for $W_k(n)$. For example, $W_2(n) \leqslant 4$ for all values of n. This is a very compact way of expressing a famous theorem of Lagrange called the 'four-square' theorem. Put into words it states that every positive number is a sum of four squares (e.g. $1 = 1 + 0 + 0 + 0$, $13 = 4 + 9 + 0 + 0$, $27 = 1 + 1 + 25 + 0$, $23 = 1 + 4 + 9 + 9$).

That $W_k(n)$ has an upper bound for any fixed k was first suspected by Waring in his book *Meditationes Algebraicae* in 1770 when he claimed that every positive number is a sum of 4 squares, or 9 cubes, or 19 fourth powers and so on. The proof of this assertion however had to wait until the early part of this century for the genius of Hilbert to discover.

We shall denote the largest value of $W_k(n)$, for fixed k, by W_k. Then $W_2 = 4$ since the four-square theorem shows that $W_2 \leqslant 4$ and the fact that $W_2(23) = 4$ shows that $W_2 \geqslant 4$. The values of W_3 and W_4 were given correctly by Waring: $W_3 = 9$ and $W_4 = 19$.

We shall now prove that $W_3(n)$ has an upper bound by showing that $W_3(n) \leqslant 13$ for all n. A more sophisticated proof is needed to improve this to $W_3(n) \leqslant 9$.

The key equation in this proof is

$$\sum_{i=1}^{4} [(m + x_i)^3 + (m - x_i)^3] = 8m^3 + 6m \left(\sum_{i=1}^{4} x_i^2 \right). \tag{1}$$

Verify this by expanding each side of the equation. Now any positive integer r can be expressed in the form

$$r = \sum_{i=1}^{4} x_i^2$$

where each $x_i \geqslant 0$. (This is just the four-square theorem.) So any number of the form $8m^3 + 6mr$ (where $r \geqslant 0$) can be expressed as a sum of 8 cubes, the left-hand side of equation (1). Since we want all such cubes to be greater than or equal to zero, we had better insist that each $x_i \leqslant m$.

Suppose that we could find integers n_0 and m such that

$$n - n_0 \equiv 8m^3 \pmod{6m} \tag{2}$$

and
$$8m^3 \leqslant n - n_0 \leqslant 14m^3. \tag{3}$$

Then $n - n_0 = 8m^3 + 6mr$ for some r (by (2)) and $0 \leqslant r \leqslant m^2$ (by (3)). Since each $x_i^2 \leqslant r$, we would have each $x_i \leqslant m$ as required. We could then conclude that $n - n_0$ is a sum of 8 non-negative cubes. If we can choose the n_0 so that it is a sum of 5 non-negative cubes, we would have proved the theorem. Here is how we can choose m and n_0.

Let m be the largest positive integer such that

$$11m^3 \leqslant n \qquad (4)$$

Then
$$n < 11(m + 1)^3. \qquad (5)$$

Now let t be the remainder on dividing $n - 4m^3$ by 6. Then

$$n \equiv 4m^3 + t\,(\mathrm{mod}\ 6) \text{ and } 0 \leqslant t \leqslant 5. \qquad (6)$$

Since at least one of $t - 1$, t, $t + 1$ is a multiple of 3 and at least one is even, $(t - 1)\, t\, (t + 1) \equiv 0\,(\mathrm{mod}\ 6)$ whence $t^3 \equiv t\,(\mathrm{mod}\ 6)$. Thus

$$n \equiv 4m^3 + t^3\ (\mathrm{mod}\ 6) \qquad (7)$$

and so
$$n = 4m^3 + t^3 + 6q \text{ for some } q \qquad (8)$$

Define x by

$$x \equiv q\ (\mathrm{mod}\ m),\ 1 \leqslant x \leqslant m. \qquad (9)$$

Finally, define n_0 by

$$n_0 = (x + 1)^3 + (x - 1)^3 + 2(m - x)^3 + t^3 \qquad (10)$$
$$= 6x + 2m^3 - 6mx\,(m - x) + t^3. \qquad (11)$$

Now n_0 is clearly a sum of 5 non-negative cubes. Since $n - n_0 - 8m^3 = -6m^3 + 6q - 6x + 6mx\,(m - x) \equiv 0\ (\mathrm{mod}\ 6m)$, equation (2) holds for this choice of n_0 and m. All that remains is the inequality (3).

From (9), (10) and (11),

$$0 \leqslant n_0 \leqslant 6m + 2m^3 + 125 \qquad (12)$$

and from (4), (5),

$$11m^3 \leqslant n < 11(m + 1)^3. \qquad (13)$$

Combining (12), (13) we obtain

$$9m^3 - 6m - 125 \leqslant n - n_0 < 11(m + 1)^3. \qquad (14)$$

Comparing (14) with (3) we can see that all would be well if $m^3 \geqslant 6m + 125$ and $11(m + 1)^3 \leqslant 14m^3$. Now while these do not hold for small values of m, they are valid if $m \geqslant 11$. So if $n \geqslant 11^4 = 14641$, $m \geqslant 11$ and the above argument shows that n is a sum of 13 non-negative cubes. To complete the proof we need only check the values of n up to 14640.

Suppose $n \leqslant 14640$ and let k_1^3 be the largest cube such that k_1^3

$\leqslant n$, let $k_2{}^3$ be the largest cube such that $k_2{}^3 \leqslant n - k_1{}^3$, and so on. Since the difference between the successive cubes k^3 and $(k+1)^3$ is $3k^2 + 3k + 1$, $n - k_1{}^3 \leqslant 3k_1{}^2 + 3k_1$. Now $25^3 > 14640$ and so $k_1 \leqslant 24$ whence $n - k_1{}^3 \leqslant 1800$.

Since $13^3 > 1800$, $k_2 \leqslant 12$ whence $n - k_1{}^3 - k_2{}^3 \leqslant 3k_2 + 3k_2 \leqslant 468$.
Since $8^3 > 468$, $k_3 \leqslant 7$ whence $n - k_1{}^3 - k_2{}^3 - k_3{}^3 \leqslant 168$.
Since $6^3 > 168$, $k_4 \leqslant 5$ whence $n - k_1{}^3 - k_2{}^3 - k_3{}^3 - k_4{}^3 \leqslant 90$.

We may easily check directly that every number up to 90 is a sum of 9 non-negative cubes. Hence n is a sum of 13 non-negative cubes.

EXERCISE SET 16

1. 150 can be expressed as a sum of squares in eight different ways. How many of them can you find? What is $W_2(150)$?

2. Verify that $W_3(n) \leqslant 9$ for all n up to 90.

3. Express 88031 as a sum of 13 non-negative cubes, 12 of which are different.
   [HINT: Put $n = 88031$ in the proof that $W_3(n) \leqslant 13$ and find the corresponding values of m, t, q, x, n_0 and r.]

4. Express 8281 as a sum of 13 non-negative cubes in each of the following ways:
   (i) Follow through that part of the proof of $W_3(n) \leqslant 13$ that applies to $n \geqslant 14641$, with $n = 8281$. Does the fact that $8281 < 14641$ raise any problems?
   (ii) Follow through that part of the proof of $W_3(n) \leqslant 13$ that applies to $n \leqslant 14640$, with $n = 8281$.

   Can you express 8281 as a sum of 13 *different* positive cubes?

5. (Harder) Prove that every positive integer is the sum of 5 cubes (where we allow some to be negative) using the identity

$$(x + 1)^3 + (x - 1)^3 - 2x^3 = 6x.$$

Solutions to the Exercises

EXERCISE SET 1

1. (b), (d), (e), (f), (i) and (j) are true; the others are false.

    [Note that 0 divides 0 since $0 = 0 \times 1$ although $0 \div 0$ is not defined.]

2. $D(15) = \{1, 3, 5, 15\}$; $D(16) = \{1, 2, 4, 8, 16\}$; $D(18) = \{1, 2, 3, 6, 9, 18\}$.

3. 17, 37, 47, 67 and 97 are primes — the others are not.

4. The claim is false since for $n = 41$, $n^2 + n + 41$ is clearly a multiple of 41. This is also the case for $n = 40$. The remarkable thing is that for $n = 0, 1, 2, \ldots 39$ this formula always gives a prime number.

5. (a) 1;   (b) 5;   (c) 3;   (d) 1;   (e) 5;   (f) 4.

EXERCISE SET 2

1. 29.

2. 9.

3. Yes.

4. If d is a common divisor of m and n it must divide $m - nq$ and so be a common divisor of $m - nq$ and n. Conversely if d is a common divisor of $m - nq$ and n it must divide $(m - nq) + nq$, viz. m, and so be a common divisor of m and n.

5. 2.

EXERCISE SET 3

1. 53424.

2. 3024.

3. 37.

4. 2520.

5. If n is odd, $(n - 1, n + 1) = 2$ and so the l.c.m. of $n - 1$ and $n + 1$ is $(n^2 - 1)/2$. Now $(n^2 - 1)/2$ is even and so $[(n^2 - 1)/2, n^2 + 1] = 2$. Hence the l.c.m. of $(n^2 - 1)/2$ and $n^2 + 1$ is $(n^4 - 1)/4$. If n is even, $(n - 1, n + 1) = 1$ and the solution continues in a similar way.

EXERCISE SET 4

1. (b), (c) and (d).

2. $x = 3$ (other solutions are possible).

3. No solutions.

4. 5, 28, 51, 74 etc. are all possible solutions. [If there are x fruits in each heap and each traveller gets y fruits then $63x + 7 = 23y$. This may be expressed as a congruence equation in x as $63x \equiv -7 \pmod{23}$.]

EXERCISE SET 5

1. (a) $x \equiv 267 \pmod{1001}$;   (b) $x \equiv 610 \pmod{1001}$;
    (c) insoluble;   (d) $x \equiv 30 \pmod{143}$.

2. (a) $x \equiv 7 \pmod{17}$;   (b) 126 adults and 62 children.

3. (a) $a = 15, b = 32$;   (b) A.D. 4301

EXERCISE SET 6

1. (a) $x \equiv 21 \pmod{37}$;   (b) $x \equiv 16$ or $36 \pmod{37}$;
   (c) $x \equiv 29$ or $36 \pmod{37}$; (d) $x \equiv 0$ or $2 \pmod{37}$.

2. $x \equiv 10$ or $26 \pmod{37}$.

3. 9, 10, 11, 12 or 25, 26, 27, 28 (other solutions are possible).
   [N.B. If the numbers are $x - 1, x, x + 1, x + 2$ then $x^2 + x + 1 \equiv 0 \pmod{37}$ and we obtain the solution from exercise 2.]

EXERCISE SET 7

1. (a) ± 32;   (b) ± 15;   (c) ± 24;   (d) ± 48;   (e) ± 14.

2. $x \equiv 25$ or $34 \pmod{41}$.

3. There are none!

EXERCISE SET 8

1. (a) 1;   (b) 1;   (c) 0;   (d) 1;   (e) 0.

2. (a) 1;   (b) 1;   (c) 1.

3. NO [since $(15|29) = 1$].

4. By property 9, $(2|4k + 1) \equiv k(2k + 1) \equiv k \pmod{2}$.

5. By property 8, $(-1|q) \equiv p \pmod{2} \equiv 1 \pmod{2}$ and by properties 1 and 10, $(q|p) = (1|p) = 0$. Hence by property 7, $0 \equiv (p|q) + (-1|p) \pmod{2}$ i.e. $(p|q) = (-1|p)$.

EXERCISE SET 9

1. Check your answers with Table 2 on page 34.

2. There must be a prime divisor, p, of $n^2 + 1$ such that $p \leqslant \sqrt{(n^2 + 1)} \leqslant n + 1$, i.e. $p \leqslant n$. Now n does not divide $n^2 + 1$ so $p < n$.

3. If k is primitive then $n^2 + n + k$ is prime for $n = 0$ (i.e. k is prime), and for $n = 1$ (i.e. $k + 2$ is prime). If $k \equiv 0 \pmod{3}$ it is not prime and if $k \equiv 1 \pmod{3}$, $k + 2$ is not prime. Hence $k \equiv 2 \pmod{3}$.

4. 5, 11, 17 and 41.

5. $n! - 1$ is not divisible by any prime $\leqslant n$. But $n! - 1$ is divisible by some prime $\leqslant n! - 1$. It is therefore divisible by a prime between n and $n! - 1$.

6. Since $2^{2n} - 1 = (2^n - 1)(2^n + 1)$, $2^{2n} - 1$ is not prime unless $n = 1$.

EXERCISE SET 10

1. (a) $T(33, 28, 65)$;   (b) $T(16, 30, 34)$;   (c) $T(10, 24, 26)$.

2. Perimeter $= X + Y + Z = 2km^2 + 2kmn = 2km(m + n)$.

3. $T(3, 4, 5)$, $T(5, 12, 13)$, $T(15, 8, 17)$, $T(7, 24, 25)$, $T(9, 40, 41)$, $T(21, 20, 29)$, $T(35, 12, 37)$.

4. Twice area $= XY = 2k^2 mn (m^2 - n^2) = kn(m - n) \times$ Perimeter.

5. $T(55, 48, 73)$.

EXERCISE SET 11

1. (a) $T(17, 144, 145)$;   (b) $T(675, 52, 677)$.

2. (a) and (b).

3. (a) and (b).

4. $T(90, 56, 106)$.

5.   If the sides are x^2, y^2, z^2 (with z^2 as the hypotenuse), then $x^4 + y^4 = z^4$. But this has no positive whole-number solutions, so no such triangle exists.

EXERCISE SET 12

1.   (a) 72; (b) 44; (c) 60; (d) 46; (e) 48.

2.   35, 39, 45, 52, 56, 72, 84 and 108.

3.   3, 5, 15, 17, 51, 85, 255, 257 and 771.

4.   None.

EXERCISE SET 13

1.   (a) 284; (b) 220; (c) 1210; (d) 1184.
      [NOTE: Numbers which are equal to the sum of proper divisors of each other are called friendly or *amicable* numbers. Thus 220, 284 and 1184, 1210 are two sets of amicable pairs. Such pairs were once held to have a mystic significance and in some cases were inscribed in wedding rings.]

2.   (a) 360; (b) 120960.

3.   If $S(n)$ is prime then n must be a prime power, say $n = p^s$. If n is even, $p = 2$ and so $S(n) = 2^{s+1} - 1$. Put $r = s + 1$. Then $S(n) = 2^r - 1$ and so is a Mersenne prime.

EXERCISE SET 14

1.   (a) 3; (b) 6; (c) 10.

2.   (a) 3; (b) 82; (c) 69.

3.   Let period of $1/p$ be d. Then d divides $\phi(p) = 40k + 2 = 2q$. If $d = 1$ or 2, p divides 9 or 99 and so equals 3 or 11, neither of which satisfies the hypotheses. If $d = 2q$, $10^q \equiv -1 \pmod{p}$ and so $10^{q+1} \equiv -10 \pmod{p}$. Thus $(-10|p) = 0$. However, $(-10|p) \equiv (-1|p) + (2|p) + (5|p) \equiv 1 + 1 + (5|p) \equiv (p|5) = 1$. The only remaining possibility is thus $d = q$.

EXERCISE SET 15

1.   348450.

2.   3094.

3. n^2.

4. $n(n+1)(n^3 - 5n + 11)/6$.

5.   2892.

EXERCISE SET 16

1. $144 + 4 + 1 + 1$, $121 + 25 + 4 + 0$, $100 + 25 + 25 + 0$, $81 + 64 + 4 + 1$, $64 + 49 + 36 + 1$, $64 + 36 + 36 + 4$, $64 + 36 + 25 + 25$ and $49 + 49 + 36 + 16$; 3.

3. $88031 = 31^3 + 9^3 + 24^3 + 16^3 + 23^3 + 17^3 + 22^3 + 18^3 + 15^3 + 13^3 + 6^3 + 6^3 + 3^3$, ($n = 88031$, $m = 20$, $t = 3$, $q = 9334$, $x = 14$, $n_0 = 6031$, $r = 150$, $x_1 = 11$, $x_2 = 4$, $x_3 = 3$ and $x_4 = 2$).

4.   (i) $1^3 + 2^3 + 4^3 + 5^3 + 5^3 + 6^3 + 6^3 + 7^3 + 9^3 + 11^3 + 13^3 + 13^3$; NO, since we still have $0 < r < m^2$ in this case.

      (ii) $20^3 + 6^3 + 4^3 + 1^3 + 0^3 + \ldots + 0^3$. $8281 = (\tfrac{1}{2} \times 13 \times 14)^2 = \sum_{r=1}^{13} r^3$.

5.   The given identity shows that every multiple of 6 is a sum of 4 cubes. It remains to show that every number is a multiple of 6 plus a cube. This is obvious for numbers of the form $6x$ and $6x \pm 1$ and follows for the others from $6x \pm 2 = 6(x \mp 1) + (\pm 2)^3$, $6x + 3 = 6(x - 4) + 3^3$.